Geothermics and Geothermal Energy

Symposium held during the joint general assemblies of EGS and ESC, Budapest, August 1980

Editors: Vladimír Čermák and Ralph Haenel

with 147 figures and 19 tables in the text

Published with financial assistance of UNESCO

E. Schweizerbart'sche Verlagsbuchhandlung
(Nägele u. Obermiller) · Stuttgart · 1982

ISBN   3 510 65109-X

Design of the cover by Wolfgang M. Karrasch
Printed in Germany

Contents

8. Case Histories and Detailed Studies of Geothermal Anomalies

Preface

The outflow of heat from the Earth's interior, the terrestrial heat flow, and the temperature field at depth are determined by deep-seated tectonic processes. The knowledge of the regional heat flow patterns is thus very important in geophysics and provides a useful tool for studying crustal and lithospheric structures and for understanding the nature of their evolution. Remarkable progress has been made in heat flow studies all over the world during the past few years. The rapidly growing interest in geothermics has been further enhanced by the possibility of using the geothermal energy as an alternative energy source.

In order to communicate the most recent advances both in heat flow and applied geothermal research, to identify important new problems and to try to find their solutions, it was decided to hold a symposium "Geothermics and Geothermal Energy" during the joint General Assemblies of the European Geophysical Society and the European Seismological Commission in Budapest, August 1980. This symposium was one of the largest of the whole meeting; a total of 50 papers were presented and discussed, the symposium took two and one-half days, i.e. 27-29 August, 1980.

The topics of the symposium covered (I) the interpretation of the heat flow data in terms of the tectonic evolution, deep crustal and upper mantle structures and their correlation with other geophysical and tectonophysical information; (II) mathematical computations and modelling of the subsurface temperature fields together with their applications to various research and prospecting projects; (III) thermal conductivity measurements, and (IV) problems of geothermal exploration, production, case histories and utilization of geothermal energy. In the latter session especially the results of the complex investigations conducted at the Urach geothermal anomaly demonstrated the necessary steps in the broad approach to this problem.

The geothermal data discussed during the symposium related to practically all Europe, with supplementary information from Iceland, Soviet Asia and India. Because of their high attendance the scientists from Eastern Europe could establish contacts with the EGS community for the first time; in addition to revealing fresh information on the existing data and their interpretation, the symposium thus gave a unique impetus for future direct cooperation among many specialists from different areas.

This volume contains the paper presented at the symposium, with a few modifications. Some papers that were presented in Budapest are not included because their authors either did not submit their manuscripts or their papers were published elsewhere; however, several originally scheduled papers that could not be read at the symposium appear in this volume. The following series of papers characterize well the present stage of the geothermal investigation and interpretation of heat flow data. The individual papers were prepared by experts in the respective fields and reflect personal views on the significance and interpretation of the results obtained in different regions. No attempt was made to standardize the approach of the individual contributors. It is believed that the various methods of analyzing and interpreting may help to overcome the complexity of the problem.

In these Proceedings of the Symposium "Geothermics and Geothermal Energy" EGS/ESC Budapest 1980, the papers have been arranged in the following order: (1) General Geothermics, (2) Regional Heat Flow, (3) Subsurface Temperature Field, (4) Thermal Conductivity Measurements, (5) Mathematical Models and Calculations, (6) Geothermal Investigations and

Relations between Various Geophysical Fields, (7) Geothermal Prospecting
and Geothermal Resources Assessments, (8) Case Histories and Detailed
Studies of Geothermal Anomalies.

The convenors together with all the participants would like to express
their sincere thanks to the local organizing committee for the perfect
organization and warm reception and to all the chairmen for their activities during the symposium. We also want to thank to Dr. E. Nägele for his
offer to publish the conference material in the Schweizerbart'sche Verlagsbuchhandlung as a special issue. Especially sincere acknowledgements are
due to the Division of Earth Sciences of the UNESCO for its financial
support, without which the realisation of this book would not be possible.

Praha and Hannover, January 1982

Vladimír Čermák and Ralph Haenel
convenors of the symposium

Regional pattern of the lithospheric thickness in Europe

Vladimír Čermák

with 5 figures

Čermak, V., 1982: Regional pattern of the lithospheric thickness in Europe. - Geothermics and geothermal energy, eds. V. Čermák & R. Haenel, E. Schweizerbart'sche Verlagsbuchhandlung, Stuttgart: 1-10.

Abstract: A geothermal model of the lithosphere, in which the surface heat flow is in equilibrium with the heat flowing into the lithosphere at its base plus the heat generated by the radioactive decay within it, is proposed. The deep temperature distribution was calculated as a function of the surface heat flow. It was assumed that the top of the low velocity is determined by the depth at which the calculated temperature reaches the melting temperature; that means that the deep temperature calculation can be used for estimating the lithospheric thickness. By using the map of the surface geothermal activity in Europe, we projected the regional variations of the lithospheric thickness in Europe. In the Precambrian shield and the platform areas of northern and eastern Europe the lithosphere is more than 200 km thick, while thin lithosphere (less than 100 km) exists beneath the active tectonic units of southern and southeastern European high heat flow zones.

Author's address: Geophysical Institute, Czechoslovak Academy of Sciences, CS-141 31 Praha-Sporilov, Czechoslovakia

Introduction

Seismic velocities in the upper mantle increase with depth, but at a certain depth there is a sharp velocity decrease and a zone of low velocity can be observed. This zone, called a low velocity channel, is a general terrestrial phenomenon and characterizes the passage between the solid lithosphere and the viscous asthenosphere. Since the seismic velocities are temperature dependent, the velocity decrease obviously corresponds to the region where the existing temperature reaches the melting temperature of the upper mantle rocks. It was the idea of Pollack & Chapman (1977) to use the calculated geotherms in combination with the mantle melting relations to estimate the top of a seismic low velocity channel and to map the regional variations of the lithospheric thickness.

By means of the 18th degree spherical harmonic representation of global heat flow, the lithospheric thickness map for Europe was projected (Chapman et al. 1979) on the basis of geotherm families published earlier (Pollack & Chapman 1977). By following their approach closely but not exactly, we proposed a more refined geothermal model of the continental lithosphere and applied it to the construction of a new version of such a map.

In contrast to the original work, a more detailed vertical distribution of the coefficient of thermal conductivity, especially its temperature dependence, based on numerous experimental data, was used. For purposes of the present paper, the heat production in the lower crust is proportional to the upper crustal radioactivity, i.e. to the surface heat flow, as compared with the constant value used in the lower crust by the above authors. Since the actual crustal thickness affects the total radiogenic contribution, i.e. also the computed temperatures, the surface heat flow pattern was completed by using the map of the regional variations of the crustal thickness (Čermak 1979a). Unlike Pollack & Chapman (1977), who based their calculations solely on the spherical harmonic representation,

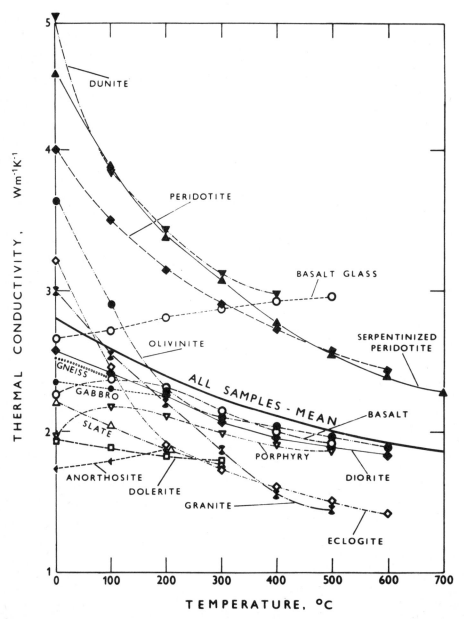

Fig. 1. Thermal conductivity of various rocks as a function of temperature.

advantage was taken of the existing heat flow map of Europe completed with the tectonic background (Čermák & Hurtig 1979). Both maps were divided into the regular 4° x 4° latitude-longitude grid elements, the mean value of heat flow and crustal thickness taken for each such element and the corresponding lithospheric thickness calculated. All individual data were then interpolated to construct the final regional pattern. By this technique the dependence on the uneven data distribution is partly removed and the resolution of the map was improved by a factor of 2.5, i.e. from less than 1100 km in the former map (Chapman et al. 1979) to about 400 km in the present one.

Thermal conductivity

Thermal conductivity of rocks generally decreases with increasing temperature and with a few exceptions, the higher the conductivity the steeper the decrease which can be observed. For typical rocks a simple formula can be used to characterize the temperature dependence of their thermal conductivity: $k = k_0 (1 + CT)^{-1}$, where k_0 is the conductivity at 0°C (i.e. approximately on the surface) and C is the parameter, the value of which is in the interval from 0 to 0.003 °C^{-1}. Fig. 1 summarizes the experimental results on the temperature dependence of the coefficient of thermal conductivity within the range 0 - 700 °C published by various authors (Birch & Clark 1940, Kawada 1964, 1966, Moiseenko et al. 1966, Moiseenko 1968, Sakvarelidze 1973, Winkler 1952). Altogether the data of 74 rock samples were collected, including 8 granites, 4 diorites, 4 gabbros, 6 dunites, 9 basalts and basalt glasses, 6 diabases, 3 eclogites, 2 olivinites, etc., among which only basalt glass and anorthosite display opposite behaviour, that is the increase of conductivity with increasing temperature. By using all the above data, we calculated the "mean" curve by the least squares method, revealing $k_0 = 2.83$ Wm^{-1}K^{-1}, C = 0.000835 ± 0.000060 °C^{-1}. Since the upper crust is dominated by granitic rocks, this curve was simplified: $k = 3/(1 + 0.0008 T)$, which is closer to granites.

The lower crust is composed of "basaltic" rocks. Basalts are relatively independent of temperature and in the interval 200-600 °C their conductivity does not differ significantly from 2 Wm^{-1}K^{-1}. This value was therefore adopted to characterize the thermal conductivity in the lower crust.

Completely different relations may exist in the upper mantle, where at higher temperatures the total thermal conductivity is determined as the sum of the lattice and radiative components: $k_T = k_L + k_R$. While the lattice conductivity decreases with temperature ($\sim T^{-1}$), the radiative conductivity, which is negligible at lower temperatures, increases with temperature and may be dominant at the upper mantle temperatures. The data published by Schatz & Simmons (1972) for a hypothetical dunite mantle material were used for further calculations and for simplicity approximated also by the above formula $k = k_0 (1 + CT)^{-1}$, in which $k_0 = 2.5$ Wm^{-1}K^{-1} and C = - 0.00025 °C^{-1} (negative), see Fig. 2.

Heat Production

The heat which is released by the radioactive decay is the main source of the heat contribution to the heat flow within the lithosphere. Its amount decreases with depth, since the radioactivity of rocks decreases with their increasing basicity. The vertical distribution of heat sources is not known by direct measurements, and it must therefore be estimated from other arguments.

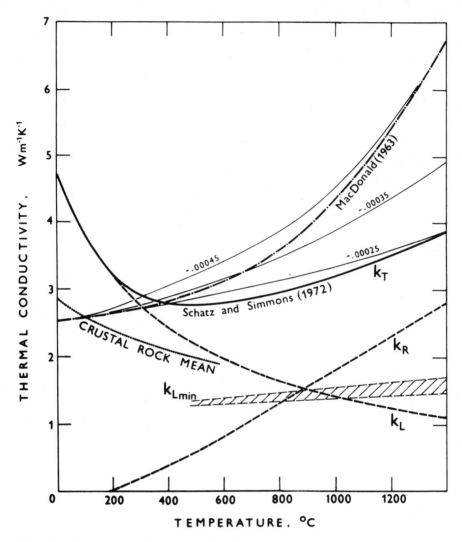

Fig. 2. Thermal conductivity in the crust and the upper mantle as a function of temperature.

The heat production in the rocks near the surface (A_0) is related to the observed surface heat flow (Q) by a simple linear relationship (Birch et al. 1968): $Q = q_0 + DA_0$, in which q_0 is the reduced heat flow (the heat flow from below a certain depth) and D is the parameter which characterizes the vertical source distribution. Many heat production distributions $A(z)$ can satisfy the above relationship; however, only two are generally mentioned.

- The step model (Roy et al. 1968/ represents the earth's crust as composed of a series of blocks of vertically constant, but horizontally varying radioactivity, with uniform depth D and uniform heat flow q_0 from below this depth.

- The exponential model of Lachenbruch (1968): $A(z) = A_0 \exp(-z/D)$, maintains the linear relationship even through the differential erosion and
  corresponds better to the geochemical principles of element redistribu-
  tion in cooling magma.

A combination of both these models can be used in the following calculation of the deep temperature distribution. In addition to the above
relation between surface heat flow and near-surface heat production,
another experimental relation between the mean surface heat flow $\bar{Q}$ and the
reduced heat flow q_0 within the corresponding heat flow province (Pollack
& Chapman 1977) was utilized: $q_0 = 0.6\ \bar{Q}$.

Let us generalize both experimental relationships and suppose that
within a certain surface area the mean heat production $\bar{A}_0$ in the surface
layer of thickness D is related to the mean surface heat flow: $\bar{A}_0 =$
$(\bar{Q} - q_0)/D$, which can be further adapted: $\bar{A}_0 = 0.4\ \bar{Q}\ /D$.

For exponentially decreasing radioactivity with depth, the mean heat
production $\bar{A}_0$ in the layer bounded by depths z_1 and z_2 is given by:

$$\bar{A}_o = \int_{z_1}^{z_2} A_{o1}\ \exp(-z/D)\ dz\ ,$$

where A_{o1} is the heat production at depth z_1. For $z_1 = 0$ and $z_2 = D$ it
follows that $A_{o1} = \bar{A}_o\ /\ (1-e^{-1})$ and $A_{o2} = A_{o1}\ e^{-1}$ is the heat production
at depth $z_2 = D$.

For the observed interval of surface heat flow values, 30 to 100 mWm^{-2},
and D = 10 km, the interval of $\bar{A}_o$ amounts to 1.2 to 4 μWm^{-3}; the interval
of A_{o1} is 1.9 to 6.3 μWm^{-3} and the interval of A_{o2} is 0.7 to 2.3 μWm^{-3},
which corresponds relatively well to the real conditions.

The radioactivity of the lower crust is even less known, but to complete
the model, the exponential function $A(z) = A_{o2} \exp(-z/D_2)$ was applied with

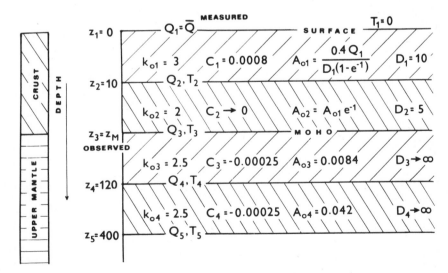

Fig. 3. Schematic model and parameters used for the deep temperature
calculation; z – depth (km), Q – heat flow (mWm^{-2}), T – temperature ($^{\circ}$C),
k – thermal conductivity (Wm^{-1}K^{-1}), C – coefficient ($^{\circ}$C^{-1}), A – heat production (μWm^{-3}), D – logarithmic decrement (km).

$D_2 = 5$ km. The mean heat production in the lower crust for the above values of A_{o2} is in the interval of 0.2 to 0.7 μWm^{-3} (for a thin crust 25 km in thickness) and in the interval of 0.1 to 0.3 μWm^{-3} (for a thick crust 45 km in thickness). The heat production at the crustal base varies over a broad interval from 0.0006 to 0.11 μWm^{-3}, according to the surface heat flow and the actual crustal thickness. The lower crust thus adds about 10 % to the observed surface heat flow and the Moho heat flow varies from about 15 mWm^{-2} in the areas of ancient shields up to 50 mWm^{-2} in the hyperthermal basins of high surface heat flow. For the heat production distribution in the upper mantle the older model by Clark & Ringwood (1964) was used with slight depth scale modifications, as proposed by Pollack & Chapman (1977). The uppermost part of the mantle down to a depth of 120 km is characterized by the depleted dunite-peridotite composition ($A_{o3} = 0.0084$ μWm^{-3}) and is underlain by the layer of primitive pyrolite ($A_{o4} = 0.042$ μWm^{-3}). Pollack & Chapman (1977) used the model by Sclater & Francheteau (1970), which gives a value of the upper mantle radioactivity which is a bit too high, and thus temperatures below 120 km which are too low and can hardly explain the conditions below the zones of low surface heat flow (< 40 mWm^{-2}). All the parameters used are schematically shown in Fig. 3.

Deep temperatures

The model of the two-layered crust and the two-layered upper mantle with temperature-dependent thermal conductivity distribution $k = k_o (1+CT)^{-1}$ and the exponential heat production $A(z) = A_o \exp(-z/D)$ gives the following steady-state one-dimensional solution $T(z)$ of the heat conductivity equation in each layer:

$$\frac{1}{C_i} \lg \left[1+C_i T(z)\right] = \frac{1}{C_i} \lg(1+C_i T_i) + Q_i \ (z-z_i) \ / \ k_{oi} - $$

$$- \frac{A_{oi} D_i}{k_{oi}} \left[D_i \exp(-\frac{z-z_i}{D_i} + z - z_i - D_i\right],$$

with

$$Q_{i+1} = Q_i - \Delta Q_i \ ,$$

where

$$\Delta Q_i = A_{oi} D_i \left[1 - \exp(-\frac{z_{i+1} - z_i}{D_i})\right].$$

In the above solution T_i and/or Q_i are the temperature and the heat flow, respectively, at the depth z_i; k_{oi}, C_i, A_{oi}, D_i are the corresponding parameters in the i-th layer, bounded by depths z_i and z_{i+1}. For $k \neq k(T)$ and/or $A \neq A(z)$ (i.e. constant within a certain layer), the solution remains valid, if respectively, $C_i \rightarrow 0$ and/or $D_i \rightarrow \infty$ in the respective layer.

The calculated temperature-depth curves for the standard crustal thickness $z_3 = z_M = 35$ km are shown in Fig. 4 together with the melting relations adopted after Chapman et al. (1979). For the evaluation of the lithospheric thickness the intersection of the temperature curve with the

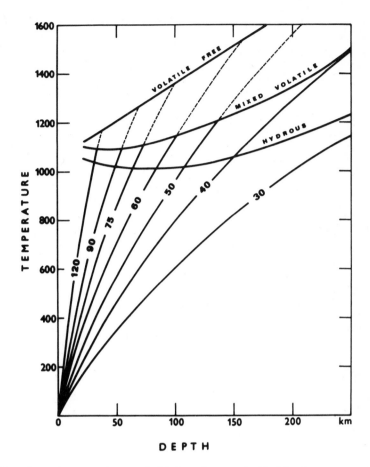

Fig. 4. Geotherm family for the continental lithosphere, individual curves
are labelled in surface heat flow (mWm^{-2}), melting relations after Chapman et al. (1979).

"mixed volatile" mantle solidus was used. Since the calculated geotherms
depend slightly on the actual depth of the Moho (z_3), the mean crustal
thickness in each grid element was taken and incorporated into the model
calculation. For a thicker crust higher upper mantle temperatures are
obtained and for a thinner crust lower temperatures are found. The actual
fan of geotherms is thus narrower in comparison with that for the standard
crustal thickness of 35 km, because of the observed negative relationship
between the surface heat flow and the crustal thickness (Čermák 1979b).

Map of lithospheric thickness

The derived map of the regional variations of the lithospheric thickness
in Europe is shown in Fig. 5. The projected thickness of the lithosphere
below the European region varies from less than 50 km to more than 200 km.
The thick lithosphere corresponds to both shields and a greater part of
the East-European Platform, i.e. to the oldest and the most stable tectonic
regions forming the European craton, and to the eastern part of the Medi-

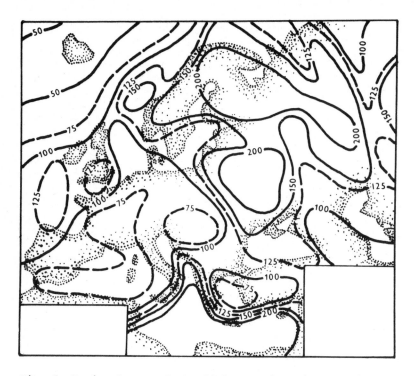

Fig. 5. Regional map of the lithospheric thickness in Europe.

terranean Sea (part of the stable African plate). The thinnest lithosphere exists in the North Atlantic along the axis of the Mid-Atlantic ridge and in the young Alpine-Carpathian belt of Neo-Europe. Thin lithosphere also characterizes the western part of the Mediterranean Sea, particularly the Tyrrhenian Sea. The lithospheric thickness thus corresponds directly to the intensity of the deep-seated processes in the upper mantle and is closely connected with the age of the crustal stabilization in its tectonic evolution.

The lithospheric thickness pattern thus obtained seems to correlate well with Adám's (1978) magnetotelluric data on the position of the high conductivity layer, which is supposed to be the upper part of the asthenosphere; its depth varies from less than 60 km below the areas of high heat flow (Pannonian Basin, South Caspian Depression) to more than 200 km below the stable platform areas of very low surface geothermal activity. Moreover, seismological data on the Rayleigh wave dispersion (Calcagnile & Panza 1980, Calcagnile et al. 1980, Mueller - personal communication) correspond fairly well to the present results, even though it is possible that the actual lithospheric thickness beneath the hyper-thermal regions, such as the Upper Rhine Graben, can be as small as 30-40 km.

References

Adám, A., 1978: Geothermal effects in the formation of electrically conducting zones and temperature distribution in the earth. - Phys. Earth Planet. Int. 17: 21-28.

Birch, F. & Clark, H., 1940: The thermal conductivity of rocks and its
    dependence on temperature and composition. - Am. J. Sci. 238: 529-558.
Birch, F., Roy, R.F. & Decker, E.R., 1968: Heat flow and thermal history
    in New York and New England. - In: E.Zen, W.S.White, J.B.Hadley &
    J.B.Thompson, jr. (Eds.), Studies of Appalachian Geology: Northern and
    Maritime. - Interscience, New York, pp. 437-451.
Calcagnile, G. & Panza, G.F., 1980: The lithosphere-asthenosphere system
    in the Italian area. - In: A.Udias & J.Channel (Eds.), Evolution and
    Tectonics of the Western Mediterranean and Surrounding Areas. - Inst.
    Geografico Nacional, Spec. publ. 201, Madrid, pp. 225-230.
Calcagnile, G., D'Ingeo, F. & Panza, G.F., 1980: The lithosphere in South-
    eastern Europe: Preliminary results. - In: A.Udias & J.Channel (Eds.),
    Evolution and Tectonics of the Western Mediterranean and Surrounding
    Areas. - Inst. Geografico Nacional, Spec. Publ. 201, Madrid, pp. 231-
    236.
Čermák, V., 1979a: Preliminary Map of the Crustal Thickness in Europe,
    1:5,000,000. - Geophys. Inst. Czechosl. Acad. Sci., Praha (unpublished).
— , 1979b: Heat flow map of Europe. - In: V.Čermák & L.Rybach (Eds.),
    Terrestrial Heat Flow in Europe. - Springer Verlag, Berlin, Heidelberg,
    New York, pp. 3-40.
Čermák, V. & Hurtig, E., 1979: Heat Flow Map of Europe. - In: V.Čermák &
    L.Rybach (Eds.), Terrestrial Heat Flow in Europe. - Springer Verlag,
    Berlin, Heidelberg, New York, enclosure.
Chapman, D.S., Pollack, H.N. & Čermák, V., 1979: Global heat flow with
    special reference to the region of Europe. - In: V.Čermák & L.Rybach
    (Eds.), Terrestrial Heat Flow in Europe. - Springer Verlag, Berlin,
    Heidelberg, New York, pp. 41-48.
Clark, S.P., Ringwood, A.E., 1964: Density distribution and constitution
    of the mantle. - Rev. Geophys. Space Phys. 2: 35-88.
Kawada, K., 1964: Studies of the thermal state of the earth - the 15th
    paper: Variation of the thermal conductivity of rocks, part 1. - Bull.
    Earthquake Res.Inst. Tokyo Univ. 42: 631-647.
— , 1966: Studies of the thermal state of the earth - the 17th paper:
    Variation of the thermal conductivity of rocks, part 2. - Bull. Earth-
    quake Res. Inst. Tokyo Univ. 44: 1071-1091.
Lachenbruch, A.H., 1968: Preliminary geothermal model of the Sierra
    Nevada. - J. Geophys. Res. 73: 6977-6989.
Moiseenko, U.I., Sokolova, Z.A. & Kutolin, V.A., 1966: Teploprovodnost
    eklogita i dolerita pri vysokikh temperaturakh. - Dokl. Akad. Nauk SSSR
    173: 163-165. (Russ.)
Moiseenko, U.I., 1968: Wärmeleitfähigkeit der Gesteine bei hohen
    Temperaturen. - Freiberger Forschh. C 238: 89-94.
Pollack, H.N. & Chapman, D.S., 1977: On the regional variation of heat
    flow, geotherms, and lithosphere thickness. - Tectonophysics 38: 279-
    296.
Roy, R.F., Blackwell, D.D. & Birch, F., 1968: Heat generation of plutonic
    rocks and continental heat flow provinces. - Earth Planet. Sci. Lett. 5:
    1-12.
Sakvarelidze, E.A., 1973: Teplofizicheskiye svoystva gornikh porod v
    intervale temperatur 20-500 ºC. - In: V.I.Vlodavec & E.A.Lubimova
    (Eds.), Teploviye potoki iz kori i verkhney mantiyi. - Verkhnaya
    mantiya No. 12, Nauka, Moscow, pp. 58-77.
Schatz, J.F. & Simmons, G., 1972: Thermal conductivity of earth materials
    at high temperatures. - J. Geophys. Res. 77: 6866-6983.

Sclater, J.G. & Francheteau, J., 1970: The implication of terrestrial heat flow observations on current tectonic and geochemical models of the crust and upper mantle of the earth. - Geophys. J. R. Astron. Soc. 20: 509-542.

Winkler, H.G., 1952: Wärmeleitfähigkeit der Gesteine. - In: Landolt-Börnstein, Zahlenwerte und Funktionen, Bd. III, Astronomie und Geophysik. - Springer Verlag, Berlin, Wien, Heidelberg, p. 324.

Global thermal regime of the lithosphere with its different permeability

E.A.Lubimova and E.I.Suetnova

with 2 figures

Lubimova, E.A. & Suetnova, E.I., 1982: Global thermal regime of the lithosphere with its different permeability. - Geothermics and geothermal energy, eds. V.Čermák & R.Haenel, E. Schweizerbart'sche Verlagsbuchhandlung, Stuttgart: 11-16.

Abstract: Recently, spherical harmonic analyses of the terrestrial heat flow data, taking into account the evidence on the different permeability of the individual tectonic units have been used for the prediction of deep temperature isolines beneath the earth's surface. High permeability in the earth's crust is expected to occur in the rifting and ridge systems, fault zones, in the regions of extension of the lithosphere, areas of hydrothermal circulation, etc. High permeability of the lithosphere means an increase of the effective thermal conductivity coefficient. The highest value of this coefficient for oceanic ridges is assumed to be equal to $3.8\ Wm^{-1}K^{-1}$. The results of the numerical investigation of deep temperature isolines are given in two maps, drawn for the depths of 50 and 25 km, which generally correspond to the Moho and/or to the upper and lower crust boundaries for continents, respectively, and to the upper mantle for oceans.
Temperatures range from 300 °C to 900 °C at a depth of 25 km and from 500 °C to 1500 °C at a depth of 50 km. It is clear that the Moho discontinuity is not the isothermal level. Consequently, the temperatures calculated for Europe are relatively low beneath East Europe and higher beneath West Europe.

Authors' address: O. Yu. Schmidt Institute of Physics of the Earth, B. Gruzinskaya, Moscow 123242, UdSSR

Introduction

The physical and tectonic states of the earth's crust and upper mantle are determined mainly by the pattern of the terrestrial heat flow. The results of the recent spherical harmonic analysis of heat flow data are used in the present paper for constructing the maps of the geotherms within the earth's lithosphere at specific depths, that is, near the boundaries between the upper and the lower crust and between the crust and the mantle. The behaviour of the most important parameters of the physical state inside the earth is assumed to depend on the regional heat flow variations and deep temperatures.

Recently the number of heat flow measurements has exceeded 7000. The average values of heat flow through the land and the ocean floor were found to be almost the same (Lee & Uyeda 1965, Horai & Simmons 1969, Chapman & Pollack 1977). Statistically the value of heat flow decreases from the younger orogenic units to the older ones (Polyak & Smirnov 1968, Sclater & Francheteau 1970). Several models of the vertical distribution

of radiogenic heat sources have been suggested; of these the most discussed are the step, the linear and the exponential models (Lachenbruch
1970).

The map of heat flow isolines was prepared by Chapman & Pollack (1975)
(making use of 829 mean values in 5 x 5° elements of the globe) on the
basis of the spherical harmonic analysis up to 12th order, supplementing
the actual data set by predicted values according to the heat flow-age
relationship. Suetnova (1979) reported a spherical harmonic representation
of surface heat flow using 1006 mean values in 5° x 5° elements on the
globe. The recent maps indicate a correlation with global tectonics: mid-
oceanic ridges are regions of anomalously high heat flow just as the
marginal seas of the Pacific Ocean, the East Pacific Rise and orogenic
areas of the land, while the Precambrian shields and platforms and the
oceanic basins are low heat flow regions.

The above maps have proved to be important for correlating the geothermal and gravitational fields. Hamza (1979) stated the heat flow-age
relationship to be valid for the continental mantle heat flow as well.

Recently attention has been paid to the maps of the mantle heat flow
isolines and the isolines of the lithospheric thickness (Pollack & Chapman
1977, Chapman & Pollack 1977). Their maps were prepared under the
assumption that the thermal conductivity coefficient is laterally constant.
Different permeability of the crust was not taken into consideration. It
is evident that high permeability of the lithosphere should be closely
connected with fault zones such as rift systems, ridges, regions of hydrothermal activity and orogenic areas (Lister 1974, Lachenbruch 1979). The
object of the present paper is to study the lithospheric temperatures
while taking into account the different permeability of the lithosphere.

Data for deep geotherm calculations

The geothermal isolines calculated from the heat flow values now available
imply the determination of internal geodynamic parameters within the
earth's crust and mantle. We used the map of the heat flow isolines based
on our last version of the spherical harmonic analysis of the latest heat
flow data (Suetnova 1979). The numerical prediction of deep geotherms in
the earth's lithosphere is based on the data of this analysis of the heat
flow, taking into account the differences of the effective thermal
conductivity coefficient.

Lister (1977) showed that in the oceanic ridge newly intruded magma
was permeable to water circulation after solidification and contraction
because of fractures created by cooling. Lowell (1975) investigated the
effect of an isolated fracture loop on the heat flow distribution near a
ridge system. The dependence of the maximum surface heat flow on permeability was discussed by Fehn & Cathles (1979). In order to construct a
map of the lithospheric temperature, we have assumed that the effective
thermal conductivity coefficient increases in the areas of high permeability.

Despite the fact that surface heat flow is fairly well known, an uneven
distribution throughout the surface still exists. We obtain the mean heat
flow for each 5° x 5° element of the globe by using all observed data.
The temperature variation with depth, which depends on the thermal properties of the rocks, was calculated for each 5° x 5° element according to
the standard procedure for the solution of the heat conductivity equation
(Tikhonov & Samarsky 1972).

With the variations of the effective thermal conductivity parameter in

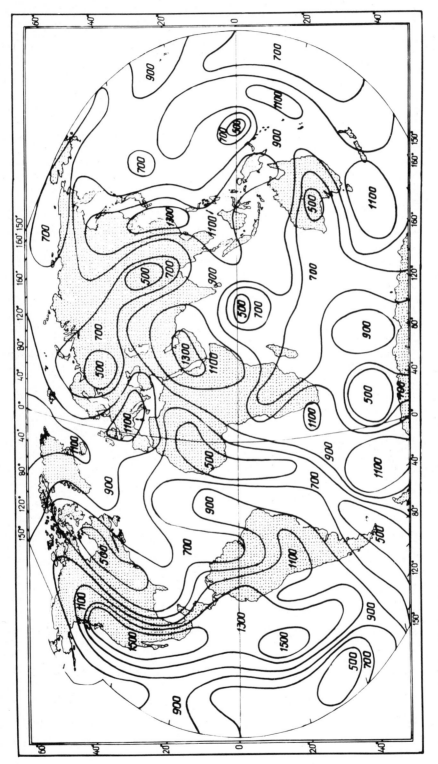

Fig. 1. Spherical harmonic representation of temperature at the depth of 50 km.

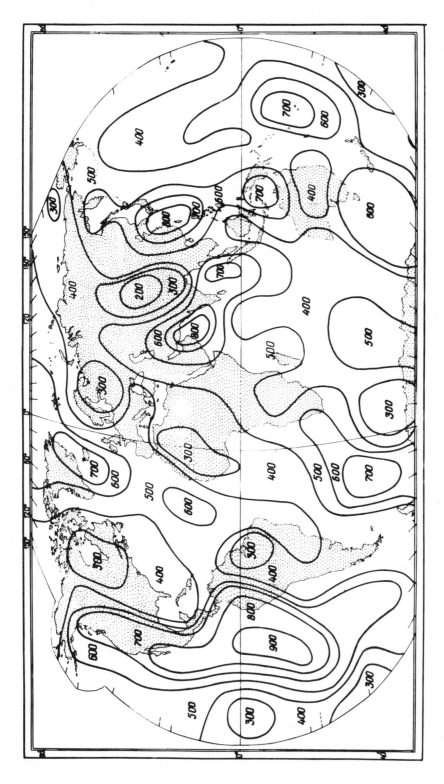

Fig. 2. Spherical harmonic representation of temperature at the depth of 25 km.

mind, we subdivide the earth's surface into five generalized types: 1. mid oceanic ridges; 2. oceanic basins; 3. stable continents; 4. transitional zones between land and oceans; 5. orogenic areas and continental rifts. Taking into account different characteristics of the tectonic units and variable permeability values for each of them, we can assume an increase of the effective conductivity coefficient for most of the faulted and fractured regions. Accordingly, the effective thermal conductivity parameter was taken as follows: for mid oceanic-ridges 3.76 $Wm^{-1}K^{-1}$, for oceanic basins 2.51 $Wm^{-1}K^{-1}$, for stable continents 2.93 $Wm^{-1}K^{-1}$, for transitional zones between land and oceans 2.72 $Wm^{-1}K^{-1}$, for orogenic areas and continental rifts 3.35 $Wm^{-1}K^{-1}$.

These values were taken according to data reported by Karakin & Lobkovskii (1980) and Fehn & Cathles (1979) who considered the level of fracturing of the crust in the zones of extension. A spherical harmonic representation of lithospheric temperatures was constructed with the use of all calculated values of deep temperatures in 5° x 5° elements on the globe.

Results

The results of the numerical calculation of deep temperature isolines for both highly permeable and nonpermeable units of the lithosphere are given in two maps in Figs. 1 and 2 for the depths of 50 and 25 km, which generally correspond to the Moho and the upper and lower crust boundaries for continents and the upper mantle for oceans.

The calculated temperatures range from 300 ^{o}C to 900 ^{o}C at a depth of 25 km and from 500 ^{o}C to 1500 ^{o}C at a depth of 50 km.

The temperature distribution at the depth of 50 km (Fig. 1) can be used as a preliminary approximation for the Moho discontinuity in terms of temperature in the thick continental crust. One can see that the Moho boundary layer cannot be taken as an isothermal surface. The range of the lateral temperature contrast reaches 1000 ^{o}C at the depth of 50 km. As pertains to the European continent, one can see the contrast between the deep temperatures of the eastern and western parts of Europe. In the Alpine zone in the western part of Europe, the temperature exceeds 1000 ^{o}C at the depth of 50 km, while in the eastern part of Europe, in the area of the East-European platform, the temperature reaches only 500 ^{o}C at the same depth. These results agree with the calculation of the European mantle temperatures performed by Cermák & Hurtig (1979). This means that the lithosphere is probably thinner in Western Europe than in Eastern Europe.

In conclusion, it is to be noted that the temperature contrast is one of the main reasons for the concentration of thermoelastic stresses followed by earthquakes.

References

Cermák, V. & Hurtig, E., 1979: The preliminary heat flow map of Europe and some of its tectonic and geophysical implications. - Pure and Appl. Geophys. 117, 1/2: 92-103.
Chapman, D.S. & Pollack, H.N., 1975: Global heat flow: a new look. - Earth Planet. Sci. Lett. 28: 23-32.
  - - , 1977: On the regional variation of heat flow, geotherms and lithospheric thickness. - Tectonophysics 38: 279-296.

Fehn, U. & Cathles, L.M., 1979: Hydrothermal convection at the slow-spreading mid-ocean ridges. - In: J.Francheteau (Ed.), Processes at Mid-Ocean Ridges. - Tectonophysics 55: 239-260.

Hamza, V.M., 1979: Variation of continental mantle heat flow with age: Possibility of discriminating between thermal models of the lithosphere. - Pure and Appl. Geophys. 117, 1/2: 65-74.

Horai, K. & Simmons, G., 1969: Spherical harmonic analysis of terrestrial heat flow. - Earth Planet. Sci. Lett. 6: 386-394.

Karakin, A.B. & Lobkovskii, L.I., 1980: Mechanics of porous two-phase visco-deformed medium and its geophysical applications. - Lett. Appl. Engng. Sci. 17: 797-805.

Lachenbruch, A.H., 1970: Crustal temperature and heat production: Implications of the linear heat flow relation. - J. Geophys. Res. 75: 3291-3300.

— , 1979: Heat flow in the Basin and Range Province and thermal effects of tectonic extension. - Pure and Appl. Geophys. 117, 1/2: 34-50.

Lee, W.H.K. & Uyeda, S., 1965: Review of heat flow data. - In: Terrestrial Heat Flow. - W.H.K.Lee (Ed.), Geophysical Monograph. Ser. N. 8. (American Geophysical Union): 87-190.

Lister, C.R.B., 1974: On the penetration of water into hot rock. - Geophys. J. R. Astr. Soc. 39: 465-509.

— , 1977: Qualitative models of spreading-centre processes including hydrothermal penetration. - In: S.Uyeda (Ed.), Subduction Zones, Mid-Ocean Ridges, Oceanic Trenches and Geodynamics. - Tectonophysics 37: 203-218.

Lowell, R.P., 1975: Circulation in fractures hot springs, and convective heat transport on mid-ocean ridge crests. - Geophys. J. R. Astron. Soc. 40: 351-365.

Pollack, H.N. & Chapman, D.S., 1977: Mantle heat flow. - Earth Planet. Sci. Lett. 34: 174-184.

Polyak, B.G. & Smirnov, Ya. B., 1968: Relationship between terrestrial heat flow and tectonics of continents. - Geotectonics 4: 205-213. (Russ.)

Sclater, J.G. & Francheteau, J., 1970: The implications of terrestrial heat flow observations on current tectonic and geochemical models of the crust and upper mantle of the earth. - Geophys. J. R. Astron. Soc. 20: 509-542.

Suetnova, E.I., 1979: The new result of spherical harmonic analysis of global heat flow data. - In: Experimental and theoretical studies of heat flow. - Nauka, Moscow: 123-137. (Russ.)

Tikhonov, A.N. & Samarsky, A.H., 1972: The equations of mathematical physics. - Nauka, Moscow. (Russ.)

Two-dimensional correlation of heat flow and crustal thickness in Europe

Vladimír Čermák and Jiří Zahradník

with 5 figures

Čermák, V. & Zahradník, J., 1982: Two-dimensional correlation of heat flow and crustal thickness in Europe. - Geothermics and geothermal energy, eds. V. Čermák & R. Haenel, E. Schweizerbart'sche Verlagsbuchhandlung, Stuttgart: 17-25.

Abstract: A simple method for the mutual correlation of the terrestrial heat flow and the earth's crustal thickness patterns (depth to the Mohorovičić discontinuity) is proposed to reveal the possible inherent relation between both parameters and then to study its regional distribution. The region studied covers most of Europe and is bounded by the meridians 8°W and 64°E and by the parallels 32°N and 71°N. For computation purposes this area was divided into a regular net of grid elements 0.5° x 0.5°. The maps showing the distribution of the correlation coefficient for different sizes of the "moving window" were drawn and the existing correlation is briefly discussed in terms of the crustal structure.

Authors' addresses: V. Čermák, Geophysical Institute, Czechoslovak Academy of Sciences, CS-141 31 Praha-Sporilov, Czechoslovakia; J. Zahradnik, Geophysical Institute, Charles University, 121 16 Praha, Czechoslovakia

Introduction

Correlation studies between the surface heat flow and the thickness of the earth's crust help in understanding the geodynamics of the lithosphere with special attention drawn to the tectonic problems. The crustal thickness may be directly related to the heat flow, since the radioactive elements which produce the heat are predominantly concentrated in the crustal rocks. Because these rocks generate a substantial part of the terrestrial heat flow, one might expect that a thick continental crust would correspond to regions of increased geothermal activity, while low heat flow might be characteristic rather of regions of a thinned crust. However, with the increasing number of results of heat flow measurements and the interpretations of explosion seismology, it is becoming obvious that there are many regions which display quite an opposite tendency. When we focus our attention on the territory of Europe, both Precambrian shields (i.e. the Baltic and the Ukrainian shields), most of the East-European Platform, the Bohemian Massif, most of Scandinavia - we can see that all have a crustal thickness of about 40 km and more and low to very low heat flow values of 35-50 mW m^{-2}. On the other hand, there are regions of high heat flow of up to 100 mW m^{-2}, such as the Pannonian Basin and Upper Rhine Graben, for instances, which display an unusually thin crust of about 25 km or less. An inverse (negative) relation between heat flow and crustal thickness based on detailed regional studies was reported by several authors in different countries and was discussed in greater detail for the

European continent by Čermák (1979a).

Even if the negative correlation, i.e. a higher heat flow being typical over a thinner crust and a low heat flow prevailing in areas with a thick to very thick crust, seems to be valid for a substantial part of Europe, there are also regions which contradict this relation (Čermák & Smithson 1981). Among these "problematic" areas one has to include above all the Alps (increased heat flow of 70-80 mW m^{-2} and thick crust of 40-50 km)

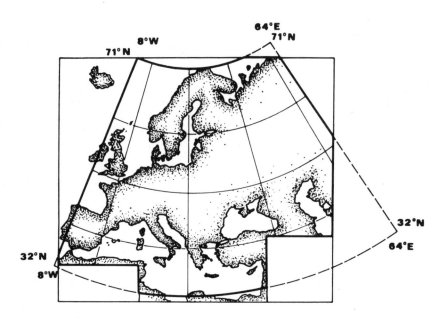

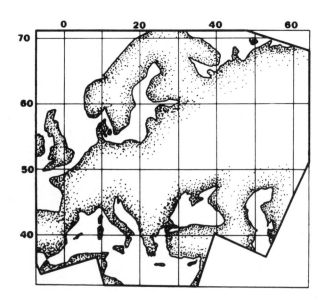

Fig. 1. Conversion from the original map format used for the heat flow and/ or crustal thickness map of Europe into the computer produced map format.

and/or the Black Sea (an area of low measured heat flow of 30-40 mW m^{-2} and subcontinental crust about 20 km thick), respectively.

Two-dimensional correlation

To be able to investigate the regional distribution of the mutual correlation between both the above parameters, i.e. to calculate their coefficient of correlation as a function of position, we proposed a simple method of two-dimensional correlation. This method facilitates the construction of the maps of the correlation coefficient and it was first tested on the territory of Czechoslovakia and its close neighbourhood (Zahradník & Čermák 1979). The present paper is an attempt to extend the area of investigation to the whole European continent.

The principle of the method is to divide the studied area into a regular net of smaller sub-areas and for each such sub-areal element to take the characteristic values of both parameters, in our case the mean heat flow and the mean crustal thickness. Both sets of data, when processed by a computer by the "moving window method" of the two-dimensional corre-

Fig. 2. Computer produced map of the heat flow distribution in Europe (simplified). Isolines are labelled in 10 mW m^{-2}.

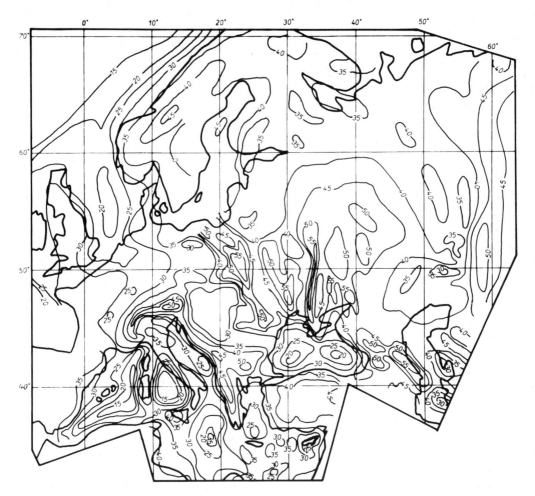

Fig. 3. Computer produced map of the crustal thickness in Europe (simplified). Isolines are labelled in km.

lation analysis, yield the regional distribution of the calculated coefficient of correlation. This distribution can then be directly related
to the tectonic setting. The degree of the division of the original field
patterns into the grid system, and the size of the moving window can be
chosen according to the density of data, their reliability and/or to the
physical proportions of the studied phenomena.

The region under study covers most of Europe and was limited by the
meridians 8°W and 64°E in the longitude and by the parallels 32°N and 71°N in
latitude. This region was covered by a grid in steps of 0.5°, i.e. it was
divided into 11 232 square elements of area 0.5° x 0.5°. The discrete
values of the heat flow and of the crustal thickness corresponding to the
individual grid points were taken from the Heat Flow Map of Europe (Čermák & Hurtig 1979) and from the Crustal Thickness Map of Europe (Čermák
1979b). Each grid value was determined in such a way that it should represent the mean value in a particular grid element. The latter map was
prepared in order to provide relevant material for the mutual correlation;
its projection and scale are the same as those used for the heat flow

pattern. The crustal thickness map of Europe was based mainly on numerous data obtained from explosion seismology published by various authors (Belyaevsky 1974, Beránek & Zounková 1977, Dachev 1980, Giese et al. 1973, Makris 1977, Morelli et al. 1967, Sellevoll 1973, Sollogub & Prosen 1973).

The conversion from the original map format to the computer produced map, and the area investigated are shown in Fig. 1. Simplified versions of the computer maps of the heat flow distribution and/or of the crustal thickness pattern in Europe are shown in Fig. 2 and Fig. 3, respectively.

The size of the moving window is characterized by K^2. i.e. by the number of grid points within it. It is obvious that a smoother correlation field can be obtained for a large window, however, for too large a window some local, geophysically important features may be suppressed and the correlation pattern loses its meaning for detailed interpretation. For areal elements of 0.5^o x 0.5^o, the corresponding dimensions of a single element vary from approximately 18 x 55 km (~1000 km^2) to 47 x 55 km (~2500 km^2) depending on the latitude. The computer procedure was repeated for values of the window parameters K = 3, 5, 7, 9, i.e. for window sizes ranging from about 54 x 165 km (the smallest window when K = 3) to about 420 x 500 km (the largest window when K = 9). Since the crustal

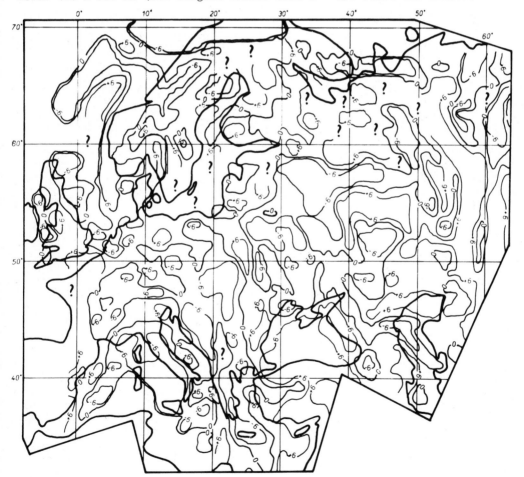

Fig. 4. Computer produced map of the regional distribution of the correlation coefficient for window size parameter K = 5 (simplified).

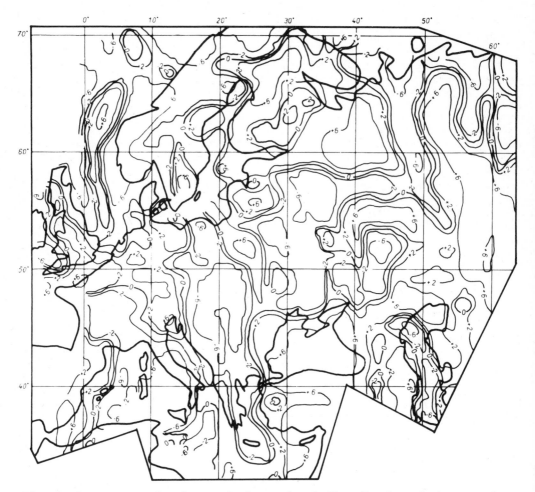

Fig. 5. Computer produced map of the regional distribution of the correlation coefficient for window size parameter K = 9 (simplified).

thickness varies from 15 to 60 km, and the lithospheric thickness from about 50 to 250 km, the window size interval mentioned above seems to cover the case well.

Correlation coefficient maps

The results obtained for the two-dimensional correlation of the heat flow versus the crustal thickness are given by the maps of the correlation coefficient for K = 5 and/or for K = 9 in Figs. 4 and 5, respectively. The data generally support the idea of the prevailing negative relation between both parameters; however, judging from the patterns obtained we do not find this conclusion so convincing as was expected, and there are several regions where an opposite tendency dominates. If the correlation pattern for K = 5 (Fig. 4) is divided into many relatively local features, it may help to identify minor anomalies in close studies. The correlation pattern for K = 9 (Fig. 5) is already smooth enough and filters out these local anomalies and shows the general pattern on a continental scale.

Both shields are characterized by a rather weak correlation. The Ukrainian shield for window size parameter K = 9 shows a correlation coefficient of -0.2 to -0.6, the negative character of the correlation seems to be generally confirmed. It is worth mentioning that the negative correlation in the Ukrainian shield and its slopes were already reported by Kutas (1979), see also the paper of Kutas in this volume. For the correlation pattern for K = 5, however, the southwestern part of the shield gave a localized opposite correlation; this local feature vanishes for a larger window, and may be caused by slightly increased radioactivity near the surface in this zone. The Baltic shield is characterized by a negative correlation in its eastern and western parts, but the central part is clearly dominated by positive correlation coefficients. This central zone of approximately SW-NE orientation divides the whole shield into two parts and roughly follows the eastern shore of the Gulf of Bothnia. As was mentioned during the EGS-ESC Meeting in Budapest in 1980 by Pirhonen et al. (1980) the new data on the crustal thickness in Finland revealed substantially higher values than the older information (Sellevoll 1973) used for the construction of the crustal thickness map of Europe employed for the present correlation. With the application of these revised data we might expect a certain improvement of the correlation pattern, in the sense of obtaining a more pronounced negative character. The Caledonian structures framing western Scandinavia reveal a strong negative character of the correlation pattern, which extends far into the North Atlantic.

Within the East-European Platform the values of the coefficient of correlation vary from -0.8 to -1.0 within the Moscow syncline and from -0.4 to -0.6 within the central parts of the Pechora, the Black Sea and Baltic synclines to high positive values of up to +0.8 within the eastern contact of the East-European Platform with the Ural Mountains, the Volga-Ural anticline and the Timan uplift zone. There seems to be a certain tectonic subdivision of the correlation pattern: the areas of depression of the basement (usually also the zones of increased geothermal activity) have negative values of the correlation coefficient, while the areas of elevation (generally of lower surface heat flow) are characterized by positive values of the coefficient of correlation. However, a deviation from this tendency is exhibited by the correlation pattern in the pre-Caspian syncline and in the zone near 40°E and 60°N (the northern part of the Moscow syncline), both being of depressed basement and both having positive values of the correlation coefficient.

Typical of the Ural Mountains range is a positive correlation over its entire length; the northwestern part of the West Siberian Platform again has the negative relation between both parameters studied.

There is an extensive zone of negative correlation in Central Europe covering the Carpathian Mountains system, its foreland and the whole intramontane depression (Pannonian Basin). This zone also includes most of the Balkan peninsula and stretches far to the west, including the Adriatic Sea, Italy, and most of the Western Mediterranean, and to the south, including Greece, the Aegean and Cretean Seas. Southern France is also a zone of negative correlation.

The Alps were reported to be a zone of positive correlation (Čermák 1979a) and this fact is confirmed by the present study, even though the coefficient of correlation is relatively small, +0.2 to +0.6. The zone of positive values of the coefficient of correlation extends from the Alps to the north covering northern France; in the form of a long narrow "arm" it stretches far to the NW across the British Isles.

Two other long, narrow zones of positive correlation are remarkable:

the first follows the axis of the North Sea (while the rest of the North Sea is characterized by negative correlation) and the other frames the western rim of the East-European Platform coinciding with the Teisseyre-Tornquist line. All the Eastern Mediterranean, the Black Sea and Turkey also create an extensive zone of positive correlation coefficient.

Conclusions

The regional correlation pattern obtained between the heat flow and the crustal thickness in Europe provides the possibility of dividing the territory studied according to the different character of the correlation. The individual sub-areas of positive and/or negative correlation can be further interpreted qualitatively and rated with respect to their tectonic structure.

Over the territory of Europe the correlation character of the two fields varies considerably, but seems to be predominantly negative. A thicker crust is thus usually characterized by a lower heat flow, while a higher heat flow belongs to regions with a thinner crust. This indicates that the terrestrial heat flow and its regional variations, regardless of the radiogenic contribution from the crust, are decisively determined by the heat flow contribution from the upper mantle. Since the temperatures deep within the crust and the upper mantle are closely related to the surface geothermal activity, this relation confirms great variations of the temperatures prevailing along the Mohorovičić discontinuity.

The region under study includes different tectonic units of contrasting crustal properties; the relation between the calculated correlation pattern is not quite clear and sometimes quite confusing, in spite of the prevailing tendency just discussed. At this stage only some very tentative conclusions can be drawn: for most depressions, regardless of their age, of the negative correlation seems to be typical, while the elevations of the basement are usually areas with positive values of the correlation coefficient, or their character is unclear. The faulted mountainous systems (such as the Alps, the Urals, and in parts of the Caucasus) are mostly areas of positive correlation, as well; the Carpathians and the Balkans, however, are of negative correlation.

The main object of this paper was to present the maps of the correlation patterns, in order to illustrate the use of the two-dimensional correlation technique. It is beyond the scope of the present work to try to give a more detailed explanation of the physical nature of the problem studied. This will certainly require a closer study in the future. More detailed knowledge of the nature of the Mohorovičić discontinuity and of the distribution of the Moho heat flow, and an assessment of its role in the tectonophysical evolution and the geological history are essential for this study.

References

Belyaevsky, N.A., 1974: Zemnaya kora v predelakh territorii SSSR. - Nedra, Moscow, 280 pp. (Russ.)
Beránek, B. & Zounková, M., 1977: Investigations of the earth's crust in Czechoslovakia using industrial blasting. - Stud. Geophys. Geod. 21: 273-280.

Čermák, V., 1979a: Heat flow in Europe. - In: V. Čermák & L. Rybach (Eds.), Terrestrial Heat Flow in Europe. - Springer Verlag, Berlin, Heidelberg, New York, pp. 3-40.
— , 1979b: Preliminary Map of Crustal Thickness in Europe, 1:5 000 000. - Geoph. Inst., Czechosl. Acad. Sci., Praha (unpublished).
Čermák, V. & Hurtig, E., 1979: Heat Flow Map of Europe. - In: V. Čermák & L. Rybach (Eds.), Terrestrial Heat Flow in Europe. - Springer Verlag, Berlin, Heidelberg, New York, enclosure.
Čermák, V. & Smithson, S.B., 1981: Correlation of heat flow and crustal thickness in Europe. - J. Geophys. Res. (in press).
Dachev, Ch., 1980: Comment on the relation of the Carpathian-Balkan-Pontian Alpine system with the block structure of the earth crust and the upper mantle. - Z. geol. Wiss. 8: 545-556.
Giese, P., Morelli, C. & Steinmetz, L., 1973: Main features of crustal structure in Western and Southern Europe based on data of explosion seismology. - Tectonophysics 20: 367-379.
Makris, J., 1977: Geophysical investigations of the Hellenides. - Hamburger Geophys. Einzelschriften, Heft 34, G.M.L.Wittenborn Söhne, Hamburg, 124 pp.
Morelli, C., Bellemo, S., Finetti, I. & Visintini, G., 1967: Preliminary depth contour maps for the Conrad and Moho discontinuities in Europe. - Boll. Geofis. Teor. Appl. 9: 142-148.
Pirhonen, S.E., Bungum, H. & Husebye, E.S., 1980: Crustal thickness in Fennoscandia as inferred from P-wave spectral ratios. - In: Programme and Abstracts, EGS-ESC Budapest 1980, Hungarian Geoph. Soc., Budapest, pp. 84.
Sellevoll, M.A., 1973: Mohorovičić discontinuity beneath Fennoscandia and adjacent parts of the Norwegian Sea and the North Sea. - Tectonophysics 20: 359-366.
Sollogub, V.B. & Prosen, D., 1973: Crustal structure of Central and South-eastern Europe by data of explosion seismology. - Tectonophysics 20: 1-33.
Zahradník, J. & Čermák, V., 1979: Two-dimensional correlation of geophysical data applied to the heat flow and earth's crust thickness patterns in Czechoslovakia. - Stud. Geophys. Geod. 23: 251-262.

Thickness of earth's crust and heat flow

R. I. Kutas

with 1 figure

Kutas, R. I., 1982: Thickness of earth's crust and heat flow. - Geothermics and geothermal energy, eds. V. Čermák & R. Haenel, E. Schweizerbart'sche Verlagsbuchhandlung, Stuttgart: 27-30.

Abstract: Within the East-European Platform and the fringing mobile belt, a few levels of heat flow values are established. Each of the levels is related to the structures of a certain age of time of activation. In the structures of the same age and type, the heat flow is directly dependent on the intensity of radiogenic heat generation in the subsurface layer of the earth's crust and inversely dependent on the crustal thickness.

Author's address: S. I. Subbotin's Institute of Geophysics, Academy of Sciences of the Ukrainian SSR, Kiev, UdSSR

Introduction

A comparison of the heat flow and the crustal thickness has been completed for the southwestern part of the USSR and adjacent territories. A large number of reports on the heat flow and crustal thickness values were used in this work (Cermák 1979, 1979a, Velinov & Petkov 1976, Kutas 1978, 1979, Kutas et al. 1978, 1979, Hurtig & Oelsner 1979, Majorowicz & Plewa 1979, Veliciu & Demetrescu 1979, Sollogub et al. 1980).

The region under study includes geological structures of various type, origin and age. It comprises the southwestern slope of the ancient East-European platform, including the Ukrainian Shield, and the mobile belt fringing the platform in the south and west. Some structures of different age of consolidation are distinguished within the belt. The Moesian plate basement and that of the Lvov and Ciscarpathian Trough were formed at the Bajkalian stage of folding (700-800 million years). The Middle-European and Scythian platform, Dobrudja, Sudeten, Bohemian Massif, etc. are Variscan structures (410-230 million years). The Alpine folding is represented by the Carpathians and the Caucasus. The tectonics of the region is essentially dependent on the processes of neotectonic activation. These are responsible for the Stavropol Rise in the Ciscaucasian region, the Black Sea and the Pannonian Basin, etc.

Heat flow and crustal Thickness

The heat flow distribution is determined mainly by the age of the structure and radiogenic heat generation in the subsurface layer of the earth's crust. High heat flow is characteristic of young structures, including the internal areas of Cenozoic geosynclines, zones of the Neogene-Quaternary rifting and zones of the tectono-magmatic activation. Intermediate heat flow values (60-75 mW/m^2) are observed in structures which were developing intensely in the Middle- and Late Paleozoic and the Early Mesozoic. A low heat flow (30-50 mW/m² is typical of the most ancient and stable blocks of the earth's crust. The mean heat flow value in the Ukrainian Shield is 36 mW/m² and it is 46 mW/m² in the ancient East-European platform. The general feature of the heat flow distribution is disturbed by tectono-magmatic activation that causes an increase of heat flow. The zones of a later activation may be marked by increased heat flows even within the ancient platform.

The crustal thickness in the above region ranges from 22 to 65 km. It depends mainly on the conditions under which evolution took place and the type of the structure. A maximum of thickness (50-65 km) is found in geosynclinal troughs irrespective of their age, a minimum (22-30 km) - in rifts and graben-like troughs. Old stabilized platforms are characterized by a thicker crust compared to young platforms.

The crustal thickness was compared with the mean heat flow values, both of which were calculated for the same elementary areas (squares of the geographic net of 1° x 1°). This has led to the conclusion that no unique correlation exists between these parameters. Only in structures of the same type and age can we find a clear inverse relationship between the

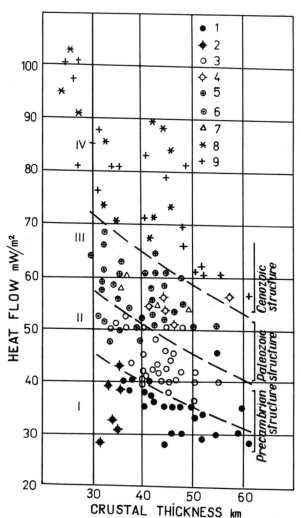

Fig. 1. Relationship between heat flow and crustal thickness. Mean values of heat flow and crustal structure are calculated for the following structures: 1) Ukrainian Shield, 2) Moesian Plate, 3) East-European Platform, 4) Activation zones of East-European Platform, 5) Variscan structures, 6) Bohemian massif, 7) Massif of Western and Eastern Carpathians, 8) Young activation zones, 9) Carpathians and Caucasus.

heat flow value and crustal thickness (Fig. 1).

In the first approximation, four types of such structures are distinguished. Type one includes the Ukrainian Shield, its slopes and the Moesian Plate. Type two includes the East-European Platform and some blocks of the Bohemian Massif, Ciscarpathian and Indolo-Kuban Alpine trough. Type three is represented by Palaeozoic structures in Central and Southern Europe, by crystalline massives of the western and southern Carpathians and some structures of the East-European Platform actively developed in the Palaeozoic (the Timan area, the dislocated zones of the Zhiguli area and Don-Medveditsa area). Type four comprises the young geo-synclinal areas and zones of tectono-magmatic activation.

Thus, within the East-European Platform and the surrounding active belt several levels of the thermal field can be distinguished. Each level corresponds to a certain age of the structure. Within this age class, the heat flow value is inversely related to the crustal thickness. It should be noted that many geological structures commonly believed to be uniform are actually heterogeneous and characterized by different levels of the thermal field.

References

Čermák, V., 1979: Review of heat flow measurements in Czechoslovakia. - In: V. Čermák & L. Rybach (Eds.), Terrestrial heat flow in Europe. - Berlin, Heidelberg, New York, Springer-Verlag, pp. 152-160.
— , 1979a: Heat flow map of Europe. - In: V. Čermák & L. Rybach (Eds.), Terrestrial heat flow in Europe. - Berlin, Heidelberg, New York, Springer-Verlag, pp. 3-40.
Hurtig, E. & Oelsner, Ch., 1979: The heat flow field on the territory of the German Democratic Republic. - In: V. Čermák & L. Rybach (Eds.), Terrestrial heat flow in Europe. - Berlin, Heidelberg, New York, Springer-Verlag, pp. 186-190.
Kutas, R.I., 1978: The heat flow field and a thermal model for the Earth's crust. - Kiev, Naukova Dumka, 140 pp.
— , 1979: The heat flow on the territory of Europe. - Geofiz. Zhurnal I: 63-73. (Russ., with Engl. Abstract)
Kutas, R.I., Sollogub, V.B. & Chekunov, A.V., 1978: The heat flow and crustal thickness in the south of the European part of the USSR. - Geofiz. sb. AN UkrSSR, Kiev, No 86:3-8. (Russ., with Engl. Abstract)
Kutas, R.I., Lubimova, E.A. & Smirnov, YA.B., 1979: Heat flow studies in the European Part of the Soviet Union. - In: V. Čermák & L. Rybach (Eds.), Terrestrial heat flow in Europe. - Berlin, Heidelberg, New York, Springer-Verlag, pp. 301-308.
Majorowicz, J. & Plewa, S., 1979: Study of heat flow in Poland with special regard to tectonophysics. - In: V.Čermák & L. Rybach (Eds.), Terrestrial heat flow in Europe. - Berlin, Heidelberg, New York, Springer-Verlag, pp. 240-252.
Sollogub, V.B., Dachev, K., Petkov, I., Posgay, K., Militzer, H., Eusberg, R., Guterch, A., Perchutz, I., Kornya, I., Constantinescu, P., Borodu-lin, M.A., Krasnopevtseva, G.V., Litvinenko, I.V., Neprochnov, Yu.P., Pomerantseva, I.V., Razinkova, M.I., Chekunov, A.V., Khalevin, I.I., Beránek, B., Prosen, D., Dragasevič, T., Andric, B., 1980: The Mohoro-vicič Discontinuity. - In: V.B. Sollogub, A. Guterch & D. Prosen (Eds.), The crustal structure of Central and South-Eastern Europe based on the results of explosion seismology. - Kiev, Naukova Dumka, pp. 123-126. (Russ.)

Veliciu, S. & Demetrescu, C., 1979: Heat flow in Romania and some relations to geological and geophysical features. - In: V. Čermák & L. Rybach (Eds.), Terrestrial heat flow in Europe. - Berlin, Heidelberg, New York, Springer-Verlag, pp. 253-260.

Velinov, T. & Petkov, I., 1976: Some results of the geothermal investigation in Bulgaria. - In: A. Adám (Ed.), Geoelectric and geothermal studies (East-Central Europe, Soviet Asia). - KAPG Geophys. Monograph. Budapest, Akadémiai Kiadó, pp. 439-442.

Geothermics and gravity in continental and oceanic rift systems*

Dietrich Werner and Hans-Gert Kahle

with 5 figures

Werner, D. & Kahle, H.-G., 1982: Geothermics and gravity in continental and oceanic rift systems. - Geothermics and geothermal energy, eds. V. Čermák & R. Haenel, E. Schweizerbart'sche Verlagsbuchhandlung, Stuttgart: 31-36.

Abstract: The Rhinegraben as a continental rift system and the Gulf of Aden as an oceanic rift system are considered. In order to interpret the gravity distribution for both cases a deep-seated geothermal anomaly within the upper mantle has to be postulated. By taking into account the temperature dependence of density it can be demonstrated that the thermally induced gravity effect plays an important role in explaining geophysical observations in rift systems.
The geothermal mantle anomaly is the result of a model simulation based on mass displacements (uprising mantle material) which are typical for rift structures.

Authors' addresses: D. Werner, Institut für Geophysik, ETH-Hönggerberg, CH-8093 Zürich; H.-G. Kahle, Institut für Geodäsie und Photogrammetrie, ETH-Hönggerberg, CH-8093 Zürich.

Introduction

In this paper the relationship between temperature field and gravity observations in rift structures is investigated. Two examples are considered: The Rhinegraben (southern Germany) as a continental rift system and the Gulf of Aden as an oceanic one. In both cases significant discrepancies are observed between seismological and gravitational findings. In order to interpret these findings it is necessary to study the problem from a geothermal point of view. We consider a deep-seated thermal anomaly reaching down to the asthenosphere. This anomaly is caused by mass displacements, i. e. uprising hot mantle material leads to an anomalous temperature field. In order to produce such a temperature field a kinematic model has been constructed.

The important point in our investigation is that the thermal effect of volume expansion on the density distribution is taken into account. By introducing these density anomalies the above mentioned discrepancy can be understood.

During the course of our study we have pursued the following procedure:

1. The kinematic model:
   A two-dimensional velocity field has been constructed in order to describe the uprising mantle flow. This mass displacement field is treated as an input parameter in our calculations, i. e. it is con-

---

* Contribution No. 310 of the Institut für Geophysik, ETH Zürich

sidered to be a given quantity.
2. The thermal model:
   On the basis of the mass displacement field the corresponding temperature field has been calculated by solving the problem of heat transport in a moving medium. This has been achieved with the aid of a finite
   difference method in space and time.
3. The density model:
   In order to calculate the lateral changes in density, the lateral
   differences in temperature have to be taken into account. The temperature distribution was transformed into a corresponding density distribution by using a linear relationship between temperature and density.
4. The gravity model:
   From the thermally induced density distribution the corresponding
   thermal gravity effect has been calculated.

The result of these calculations can be used to explain quantitatively
the discrepancy mentioned above. More details about the method are
described in two papers (Werner & Kahle 1980, Kahle & Werner 1980).

The Rhinegraben

The Rhinegraben between Frankfurt and Basel, whose formation started in
Mid-Eocene times, is the most pronounced surface expression of an extended

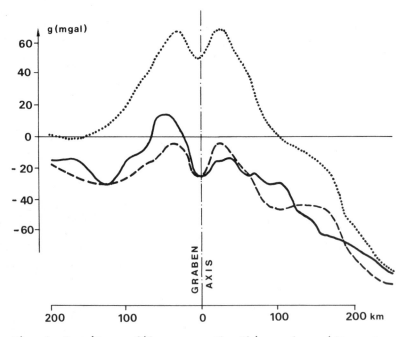

Fig. 1. Gravity profile across the Rhinegraben rift system near Strasbourg.
Dotted line: Calculated gravity anomaly based on seismic and geological
            data,
Dashed line: Calculated gravity anomaly including the thermal density
            effect taken from the temperature model in Fig. 2,
Solid line: Observed Bouguer anomaly. It can be seen that thermal density
            effect provides an explanation of the gravity discrepancy.

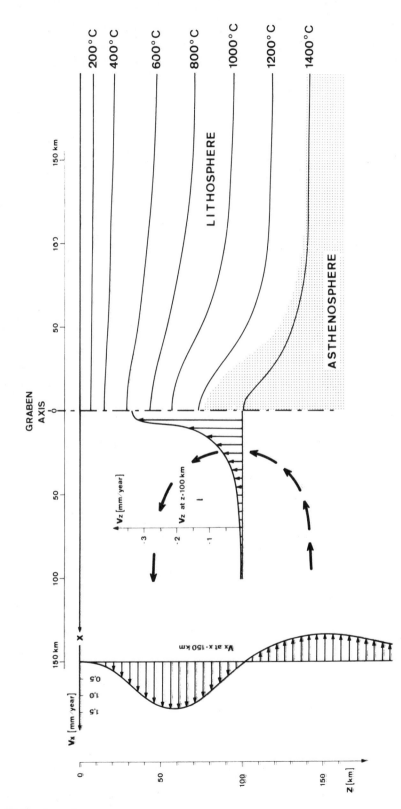

Fig. 2. Model of material and heat transport in the uppermost mantle underneath the Rhinegraben.
Left: The velocity field $v = (v_X, v_Z)$ for the upwelling mantle which produces a thermal anomaly.
Right: The temperature field after 45 Ma caused by convection and thermal conduction; a calculated boundary between lithosphere and asthenosphere (stippled area) depending on temperature and pressure is also indicated.

continental rift system traversing the crust and upper mantle of Central Europe. A nearly complete coverage of the accomplished results can be found in several proceedings of Rhinegraben symposia such as those edited by Rothe & Sauer 1967, Illies & Mueller 1970 and Illies & Fuchs 1974. The gravitational and geothermal problem outlined here has been described by Werner & Kahle 1980 and Kahle & Werner 1980.

The structure of the earth's crust and the topography of the crust-mantle boundary derived from numerous seismological results in the vicinity of the Rhinegraben is well documented and is characterized mainly by a minimum crustal depth of 24 km beneath the southern Rhinegraben (30 km crustal thickness outside the disturbed area). The density distribution in such a structure should be associated with a pronounced positive gravity anomaly of the order of 100 mgal. A comparison, however, between the expected anomaly and the observed gravity field demonstrates quite clearly that such an anomaly does not exist. Fig. 1 shows this discrepancy in a profile perpendicular to the graben at the latitude of Strasbourg: The dotted line indicates the theoretically expected anomaly, the solid line indicates the observed gravity anomalies (taken from the Bouguer maps of Germany and France).

In order to determine the "expected" anomaly, detailed three-dimensional calculations have been carried out with the following density effects taken into account:
- lateral changes in crustal thickness (the most important effect),
- the sedimentary fill of the graben proper,
- lateral changes in crustal structure,
- lateral changes in thickness of Mesozoic layers,
- the crustal structure underneath the Alps,
- the sedimentary fill of the Molasse basin.

The discrepancy shown in Fig. 1 can be eliminated by the kinematic-geothermal concept mentioned above. We assume that a mass displacement field may exist as indicated in Fig. 2 (left part). The mantle flow in our kinematic model is of the order of 1 mm/year. The period of that process is assumed to be equal to the age of the Rhinegraben (about 48 million years). Our computer programme for solving the problem of heat transport in a moving medium leads to a present day result of temperature distribution as indicated in the right part of Fig. 2. On the basis of this temperature distribution the thermally induced field of density differences and finally the thermal gravity effect, can be determined. If this thermal effect is superimposed on the "expected" gravity values (dotted line in Fig. 1), the observed Bouguer anomalies are matched almost perfectly (broken line in Fig. 1).

The Gulf of Aden

A similar problem as in the Rhinegraben occurs in the case of the Gulf of Aden, which serves as an example for an oceanic rift system. Again the observed gravity anomalies do not show a systematic gravity high in the vicinity of the Gulf of Aden. In contrast to this observation the structure of the crust and upper mantle under the gulf has been shown to be of typical oceanic character on the basis of marine geophysical results.

An attempt to estimate the theoretical gravity effect of the crust and upper mantle structure carried out by Laughton & Tramontini 1968 led to the conclusion that an anomalous low density mantle has to be invoked in this area. We interpret their proposed anomalous mantle as a deep-seated geothermal anomaly similar to that in the Rhinegraben case.

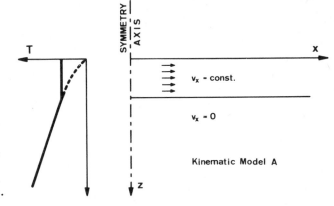

Fig. 3. A simple slab model for the sea floor spreading process in the Gulf of Aden. The temperature distribution at the symmetry axis (rift centre) is assumed to be constant with time.

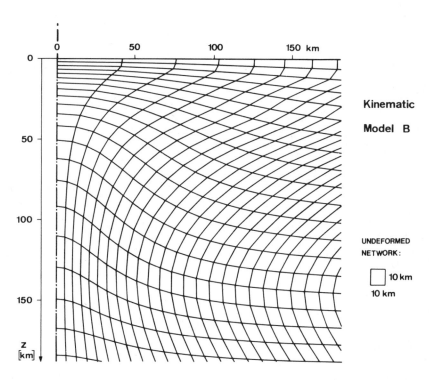

Fig. 4. A more complex displacement model for the upper mantle of the Golf of Aden. Instead of the velocity field the deformation state after 10 Ma is shown. This model is similar to that for the Rhinegraben.

Two kinematic models have been constructed. Fig. 3 shows a slab model which may be realistic with regard to the magnetic strip pattern at the sea bottom of the Gulf of Aden. However, this "model A" is probably un-suitable for describing the mantle flow. On the other hand, the model B (Fig. 4) which is very similar to our Rhinegraben model, cannot describe the geophysical findings at the ocean bottom in the Gulf of Aden.

A preliminary result is shown in Fig. 5. It can be seen that the model B is in good agreement with the observed gravity data. We are currently in the process of considering a combination of the two models shown here.

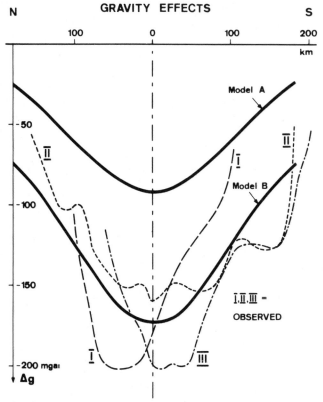

Fig. 5. Gravity profiles across the Gulf of Aden.
Broken lines: Observed gravity anomalies from various sections of the
              gulf,
Solid lines: Results of preliminary calculation based on the models
              shown in Fig. 3 and 4.

References

Illies, J.H. & Mueller, S. (Eds.), 1970: Graben problems. - Schweizerbart,
    Stuttgart, 316 pp.
Illies, J.H. & Fuchs, K. (Eds.), 1974: Approaches to taphrogenesis. -
    Schweizerbart, Stuttgart, 460 pp.
Kahle, H.-G. & Werner, D., 1980: A geophysical study of the Rhinegraben -
    II: gravity anomalies and geothermal implications. - Geophys. J. 62:
    631-647.
Laughton, A.S. & Tramontini, C., 1969: Recent studies of the crustal
    structure in the Gulf of Aden. - Tectonophysics 8: 359-375.
Rothe, J.P. & Sauer, K. (Eds.), 1967: The Rhinegraben progress report
    1967. - Abh. Geolog. Landesamt Baden-Württemberg, Heft 6, 146 pp.
Werner, D. & Kahle, H.-G., 1980: A geophysical study of the Rhinegraben -
    I: Kinematics and geothermics. - Geophys. J. 62: 617-629.

Geothermal model of the heat anomaly of the Pannonian Basin

L. Bodri and B. Bodri

with 3 figures and 1 table

Bodri, L. & Bodri, B., 1982: Geothermal model of the heat anomaly of the Pannonian Basin. - Geothermics and geothermal energy, eds. V. Čermák & R. Haenel, E. Schweizerbart'sche Verlagsbuchhandlung, Stuttgart: 37-43.

Abstract: On the basis of the solution of the steady-state, three-dimensional equation of heat conduction with variable thermal conductivity and heat generation, a geothermal model of the crust and the upper mantle beneath the Pannonian basin is constructed. The results of these calculations indicate that the surface heat flow in the Pannonian basin has an average positive anomaly of 40 mW/m^2 over the normal radiogenic flux. The temperature anomaly at the Moho-discontinuity is 500 ºC on the average and it increases with depth. The melting temperature is already attained at depths of 70-80 km. It is pointed out that some additional, non-radiogenic heating must account for the Pannonian heat anomaly.

Authors' address: Department of Geophysics, Eötvös University, 1083 Budapest, Hungary

As is well known, any heat flow value obtained in the near-surface regions of the crust is determined mainly by two factors: radioactive heat genera-tion and tectono-magmatic activity. The tectonically active zones are usually characterized by positive geothermal anomalies. The source mechanism of these anomalies, obviously produced by deep-seated processes disturbing the thermal and the dynamic equilibrium of the crust and the upper mantle, is not yet identified and understood. Indications of geo-thermal anomalies, i.e. separation of the radiogenic component from the temperature field, however, may be of great use in investigating the mechanism of extra heating in active zones.

   In this work an attempt is made to indicate the heat anomaly of the Pannonian basin, - a young sedimentary basin presumably formed by certain Late Cenozoic tectonic movements of subduction and possibly of continent-continent collision type. Though tectonically inactive at present, the Pannonian basin manifests a unique geothermal activity. The average value of the observed heat flow over the region is 85 mW/m^2, in some particular areas the surface heat flux even reaches values of 110-120 mW/m^2. Only about one-third (more exactly, 30 mW/m^2) of the mean observed heat flow can be attributed to the heat generation within the thin (on the average 26 km thick) crust, thus an average heat flux of 55 mW/m^2 is of mantle origin in the basin. If one takes a heat flow of 45 mW/m^2 as normal (the mean surface heat flow on Precambrian platforms where practically the whole observed heat flux is of radiogenic origin and is unaffected by tectonic processes), one finds that the anomaly of the heat flow in the Pannonian basin is 40 mW/m^2 on the average. Having solved the steady-state, one-dimensional equation of heat conduction in a layered medium with a structure corresponding to the average composition of the crust and the

Tab. 1. Geothermal model parameters used in calculations

Layer	Depth interval filled by the layer (km)	Thermal conductivity (W/mK)	Heat generation $(10^{-6}W/m^3)$
Sediments	0 - 2	1.8	1.0
Basement	2 - 5	2.5	1.0
Granite	5 - 18	2.1	2.1
Gabbro	18 - 26	2.3	0.4
Upper mantle	26 - 80	4.2	0.04

upper mantle beneath the Pannonian basin (the thermophysical parameters of the model are presented in Tab. 1), one can obtain the mean anomaly of deep temperature in the area of interest. The numerical results obtained by this one-dimensional approach indicate that the average anomaly of the temperature at the Moho is about 500 °C, while the corresponding normal temperature does not exceed 300-400 °C. The anomaly of temperature increases with depth; at a depth of 60 km the anomalous temperature reaches 750-800 °C with the corresponding normal temperature at about 450-500 °C. At a depth of 70 km the melting temperature beneath the Pannonian basin is attained. In these calculations an exponential decrease of heat generation in the downward direction is taken within each layer and only the lattice component of the thermal conductivity is taken into account. Thus, such a numerical estimate of the Pannonian heat flow anomaly implies that the normal radioactive heating produces only about half of the observed mean surface heat flow. With increasing depth the contribution of radiogenic

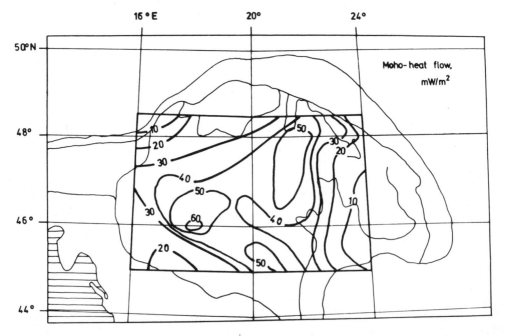

Fig. 1. Anomalous heat flow at the Moho beneath the Pannonian basin.

heating to the total heat flow decreases; at the Moho, for example, only about one-third of the total heat flux may be of radiogenic origin. There-fore the main conclusion that can be drawn from the one-dimensional numerical estimation is that some additional source of heating besides the radiogenic heating is of great importance in the establishment and mainten-ance of the heat anomaly of the Pannonian basin. The present temperature at depth in the area of interest, which can be considered as the result of cooling from the initial state of maximal heating in the past, is more than twice as high as the normal value in continental areas. The anomalous-ly high and highly stable heat flow in the region is comparable with those observed in typical active marginal basins of Early- and the Middle Tertiary.

More detailed information on heat flow anomalies can be obtained by also considering the local inhomogeneities in the spatial distribution of the observed heat flow. The spatial variation of the Pannonian heat flow anomaly was investigated on the basis of the solution of the steady-state, three-dimensional equation of thermal conduction in a layered medium which is heterogeneous both horizontally and vertically. The basic data used in these calculations are the surface heat flow (Cermák & Hurtig 1979) and the crustal structure (Sollogub et al. 1978) over the region. The area under investigation extends from $16°E$ to $24°E$ of the geographical longitude, from $45°N$ to $48.5°N$ of the geographical latitude and the lower boundary of this area is seated at a depth of 80 km.

Fig. 1 shows the distribution of the anomalous heat flow at the Moho discontinuity beneath the Pannonian basin. This heat flow can be inter-preted as a part of the observed heat flux that is produced solely by some additional, non-radiogenic source of heating. This anomalous compo-nent reaches a maximal value of 60 mW/m^2; its average value in the area of interest is 40 mW/m^2. Because of a relatively homogeneous structural composition of the crust in the Pannonian region, the heat flow pattern presented in Fig. 1 is quite similar to the distribution of the surface heat flow there. This implies that not only the generally high values of the heat flow but also a substantial share of its variation over the area of interest should be attributed to non-radiogenic heating.

An interesting feature of the thermal state of the Pannonian basin is that the inhomogeneities of the temperature field increase remarkably with depth. While the radiogenic part of the temperature field has practically a laterally uniform structure, the surfaces of equal temp-erature anomalies at medium depths exhibit a vertical variation of about ±20 km and near the bottom of the studied area this variation reaches as much as 30-40 km (see Fig. 2). It should be pointed out that the "anomaly" of temperature means temperature in addition to the "normal" value produced by radioactive heat generation alone. A temperature anomaly of 800 °C is chosen because this is the maximal possible temperature over the normal radiogenic temperature which can be taken without melting the material in the investigated area.

It can be observed in Fig. 2 that the surfaces of equal temperature anomalies fall rapidly along the meridians, approximately $22.5-23.0°E$. At the other three boundaries of the studied area (especially at the western boundary) these anomalies diminish more smoothly. Such a configura-tion of the temperature anomaly surfaces can be explained by continuous penetration of cold material into the crust and the upper mantle near the eastern boundary of the modelled region. The interpretation of deep temperature anomalies obtained here is beyond the scope of this paper; nevertheless, it is worth mentioning that the characteristic structure of these temperature anomalies strongly suggest the identification of the cold

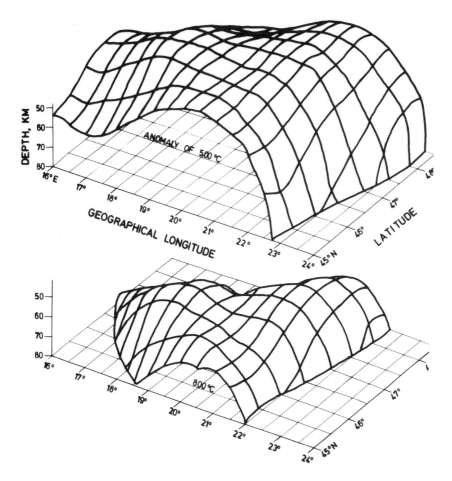

Fig. 2. Surfaces of equal temperature anomalies of 500 °C and 800 °C.

eastern zone of meridional strike with a past, at present inactive, zone of subduction.

Fig. 3 indicates the position of the zone of melting (shadowed area) beneath the Pannonian basin. The mean depth of this zone is about 75 km; in the eastern part of the basin, however, melting already occurs at a depth of 60 km. High temperature and probably partial melting under the Pannonian basin are suggested also by the elevated position of the high-conductivity layer (at depths of 60-75 km) and/or the low-velocity layer as indicated by Adám (1976) and Bisztricsány (1974), respectively. According to Posgay (1975), the upper boundary of the LVL in the Pannonian region is situated at a mean depth of 75 km and in the eastern part of the basin this boundary already appears at much smaller depths (about 57 km).

As is already known, the anomalously small thickness of the crust in the Pannonian basin is due to the thinning of its basaltic layer. It is generally accepted in relevant studies that this thinning is attributed to past melting that occurred at the bottom of the crust (subcrustal erosion). If the surface along which the initial melting occurred is considered as equivalent to the Moho-discontinuity observed at present (the latter is represented by the upper surface in Fig. 3), a very weak

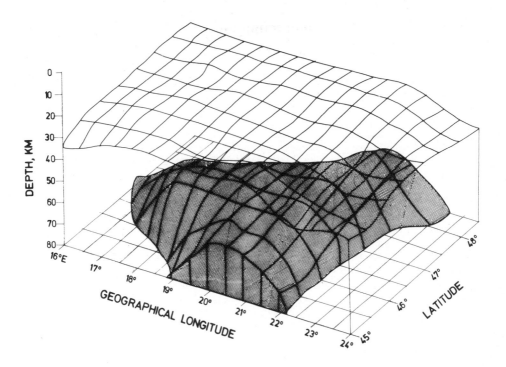

Fig. 3. Position of the zone of melting (shaded area) and the Moho-discontinuity (upper surface) in the Pannonian region.

correlation between the surfaces of melting in the past and at present can be observed. Such a weak correlation can be explained if one supposes a significant spatial variation of the thickness of the melted zone in the past.

Simple calculations of conductive cooling of initially melted layers of infinite horizontal extension, with a common upper boundary depth of 25 km and different thicknesses, indicate, for instance, that a 15 km thick layer will have been transformed completely to the crystalline state by 3 Ma from the start of cooling. For 35 km thick and 55 km thick layers the corresponding time intervals will be 9 Ma and 18 Ma, respectively. Thus, a 15 km thick layer that was melted in the Middle Miocene (the period of maximal activity of the andesite-rhyolite volcanism in the area of interest), has already been transformed completely to the crystalline state, while part of a 55 km thick layer in the depth interval from 65 km to 80 km remains melten at present, too. In the light of the above considerations it can be concluded from Fig. 3 that the region of melting had a maximal thickness in a zone of meridional strike located between the meridians of 20°E and 21°E. The location of this zone fits well with the position of the zone of presumed subduction mentioned above. Since the location of the zone (or zones) of past subduction cannot be revealed solely on the basis of geological data currently available for the Pannonian region, the investigation of geothermal anomalies, as an additional source of information, may also be of use in solving this problem.

Bodri 1976 indicated a strong negative correlation (72 %) between the variation of the surface heat flow and the depth of the Moho in the

Pannonian basin. Such a correlational relationship can be extended also to a more general case. If one constructs the curve of dependence of the lithosphere thickness on the surface heat flow for the Pannonian basin, following the method of Chapman & Pollack (1977), one finds that the thickness of the lithosphere here varies in the range of 24 km to 36 km. The correlation between the thickness of the crust and that of the lithosphere turns out again to be as strong as it was between the crustal thickness and the surface heat flow. Moreover, the corresponding regression line indicates practically a total coincidence of the crust and the lithosphere in the Pannonian region. A rise of the lithosphere obtained in the E-NE part of the region, near the suggested zone of subduction, is very compatible with certain results of numerical modelling of the tectonic flow and temperature beneath island arcs carried out by Bodri & Bodri (1978). These results also indicate significant thinning of the lithosphere above downgoing slabs.

As to the source of extra heating in the area of interest, the results of the present analysis indicate that this heating should have been of mantle origin and its intensity 2-3 times greater than that of the radioactive heat generation. The anomalous temperature field produced by the additional heat is strongly inhomogeneous. The area of maximal heating, for example, is located just near the suggested zone of penetration of cold material into the asthenosphere. It seems to be quite reasonable to attribute the source of the supposed extra heating to large scale mass flow (secondary or induced convection) that proceeded in the upper mantle. No doubt the mechanism of this heating must have been more complicated than simple upwelling of hot material (advection) from greater depths to the base of the lithosphere. At depths of 30-50 km the melting temperature is about 1200-1300 °C. In order to maintain melting here, the hot material must have been transported to these horizons from depths of 120-150 km. As indicated by the numerical studies of Bodri & Bodri (1978, 1979), in the case of intermediate velocities of subduction (1-5 cm/a) and the usual assumptions on the variation of viscosity of the asthenosphere with depth, the velocity of convective mass flow already decreases by a factor of 2 at a distance of 100 km from the descending slab. At distance of 200 km this velocity is about ten times less than that of the slab. This means that upwelling of hot material along the ascending branch of the convective flow mentioned above would need time intervals from 20 to 100 Ma. Such long time intervals obviously cannot be compatible with the presumed young age of the Pannonian basin. Besides, during the time intervals of 10^7-10^8 a, the conductive cooling of the upwelling hot material would already become significant, too. A relatively unimportant role of the advection mechanism in thermal effects of induced convection is suggested also by other model calculations (e.g., Toksöz & Hsui 1978) and by experimental evidence on the highly heated state of young island arcs.

According to geological indications, the present state of the Pannonian basin is the result of some tectonic movements that began in the Middle Oligocene - Early Miocene and the subsequent volcanism proceeded here with maximal intensity in the Middle- and Late Miocene. Therefore, one can conclude that the additional, non-radiogenic heat source was able to produce significant heating for a relatively short time in most parts of the basin. The nature of this extra heating is not yet clear. Probably its most likely mechanism is the frictional heating produced by the convective flow slipping along the base of the lithosphere.

References

Adám, A., 1976: Connection between geoelectric and geothermal parameters in the earth. - In: A. Adám (Editor-in-Chief), Geoelectric and Geothermal Studies (East-Central Europe, Soviet Asia). - KAPG Geophysical Monograph. Budapest, pp. 567-571.

Bisztricsány, E., 1974: The depth of the LVL in Europe and some adjacent regions. - Geofizikai Közlemenyek 22: 61-68.

Bodri, L., 1976: Deep temperature and heat flow in the Pannonian basin. - Ph. D. thesis, Eötvös Univ., Budapest.

Bodri, L. & Bodri, B., 1978: Numerical investigation of tectonic flow in island-arc areas. - In: M. N. Toksöz (Ed.), Numerical Modeling in Geodynamics. - Tectonophysics 50: 163-175.

— — , 1979: Flow, stress and temperature in island-arc areas. - Geophys. Astrophys. Fluid Dynamics 13: 93-105.

Čermák, V. & Hurtig, E., 1979: Heat Flow Map of Europe - Enclosure for "Terrestrial Heat Flow in Europe". - In: V. Čermák & L. Rybach (Eds.), Terrestrial Heat Flow in Europe. - Springer Verlag, Berlin-Heidelberg-New York, 328 pp.

Chapman, D. S. & Pollack, H.N., 1977: Regional geotherms and lithospheric thickness. - Geol. Soc. Am. J. 5: 265-267.

Posgay, K., 1975: Mit Reflexionsmessungen bestimmte Horizonte und Geschwindigkeitsverteilung in der Erdkruste und im Erdmantel. - Geofizikai Közlemenyek 23: 13-18.

Sollogub, V.B., Guterh, A. & Prosen, D., 1978: Structure of the Crust and Upper Mantle in Central- and East-Europe. - Naoukova Dumka, Kiev, 272 pp. (Russ.)

Toksöz, N. M. & Hsui, A., 1978: Numerical studies of back-arc convection and the formation of marginal basins. - In: M. N. Toksöz (Ed.), Numerical Modeling in Geodynamics. - Tectonophysics 50: 177-196.

Palaeogeothermics in the Ruhr Basin

G. Buntebarth, I. Koppe and M. Teichmüller

with 3 figures and 1 table

Buntebarth, G., Koppe, I. & Teichmüller, M., 1982: Palaeogeother-
mics in the Ruhr Basin. - Geothermics and geothermal energy, eds.
V. Čermák & R. Haenel, E. Schweizerbart'sche Verlagsbuchhandlung,
Stuttgart: 45-55.

Abstract: The increase in degree of coalification with depth
(coalification gradient) allows conclusions about the increase of
the former temperature of the crust with depth, i.e. on the palaeothermal gradient. Coalification data (vitrinite reflectivity, % Rm)
of coal beds from 53 deep boreholes in the mining district of the
Ruhr Basin and the adjacent Lower Rhine District were used to
calculate the geothermal gradients during the Westphalian. This
procedure is possible because the coalification is pre-orogenic in
these areas. It was completed before the Asturian folding (Westphalian D to Lower Stephanian) which caused an upheaval, and along
with the uplift a cooling of the coal beds of the Westphalian A, B
and C. The coalification gradients are higher in the Westphalian A
(mean: 0.104% Rm/100 m) than in the Westphalian B and C (mean:
0.048% Rm/100 m). This may be due to the higher level of coal rank
in the Westphalian A (1.0 - 2.3 % Rm) compared with Westphalian B
and C (0.65 - 1.3 % Rm). The higher rank of Westphalian A coals
reflects the deeper subsidence of these older strata. Another
explanation is offered by the assumption that higher temperature
gradients (mean: 79°C/km) were responsible for the higher coalification gradients in the Westphalian A, whereas the slower increase of
coalification with depth in the Westphalian B and C was caused by
lower geothermal gradients (mean: 65°C/km).
Altogether, the results of calculation of the palaeothermal regime
suggest very high temperature gradients during the Upper Carboniferous in the Ruhr Basin (60 - 80°C/km). The anomalously high thermal regime with a regional heat flow density of more than 100 mW/m^2
implies a much thinner crust within the Ruhr area during Westphalian
times than at present.

Authors' addresses: G. Buntebarth and I. Koppe, Institut für Geo-
physik, Technische Universität Clausthal, Postfach 230, D-3392
Clausthal-Zellerfeld (F. R. Germany); M. Teichmüller, Geologisches
Landesamt Nordrhein-Westfalen, de-Greiff-Straße 195, D-4150 Krefeld
(F. R. Germany)

Introduction

The Ruhr Basin represents the southern marginal part of the Subvariscan
foredeep. It is comparable with the Molasse trough north of the Alps which
represents the northern foredeep of the Alps. In these two large foredeeps
the degree of coalification is strikingly different. The degree of coalification (coal rank) is a well-known measure for the former heat exposure

of fossil organic matter (M. & R. Teichmüller 1968). Therefore different coalification patterns allow geothermal conditions to be inferred for these two foredeeps.

Whereas in the foredeep of the northern Alps, the molasse bears only lignites and sub-bituminous coals, and geothermal gradients are correspondingly low (M. & R. Teichmüller 1975), the Variscan foredeep bears bituminous coals and even anthracites. The geothermal gradient in the molasse north of the Alps is 23°C/km and less, thus corresponding with the low rank of coals and coaly matter in clastic rocks. In the molasse of the Subvariscan foredeep, i.e. the Ruhr District, the present geothermal gradient is about 30°C/km, but the present heat flow is certainly not responsible for the relatively strong coalification of the Ruhr coal seams.

Since the rank of coal does not depend only on the temperature but also on the duration of heat exposure, the question arises whether the coalification time in the Ruhr Basin was longer than that in the molasse north of the Alps. This is not the case, because the coalification in the Ruhr Basin was already completed during the latest Carboniferous (M. & R. Teichmüller 1971). Because of the Asturian orogeny the coal seams were already upfolded during the Stephanian and thus reached cooler horizons of the crust than at the time of their deepest subsidence in Westphalian times. Later, in contrast to the Carboniferous of northern Germany, the Ruhr Carboniferous did not subside deeply again, and so it did not experience a post-orogenic "post-coalification". This is why palaeogeothermal gradients of Carboniferous times may be inferred from the increase of coalification with depth, i.e. from the coalification gradients at the Ruhr.

On the basis of a comparison of coalification data and the known geothermics of the Upper Rhine Graben, R. Teichmüller (1973) estimated a mean palaeogeothermal gradient of 70°C/km for the Upper Carboniferous at the Ruhr. According to Scheffer (1965) and Čermák (1979), a close relationship exists between heat flow density and depth of the upper mantle; therefore R. Teichmüller raised the question of whether the crustal thickness was less in the Carboniferous than it is today in the Ruhr Basin and in the northern Alpine foredeep where according to Giese & Stein (1971), the Moho at present lies at a depth of 34 km (in the Alps even much deeper).

The aim of this paper is to clarify the conditions of geothermal gradients, heat flow density and crustal thickness during Carboniferous times in the Ruhr area by evaluating coalification data with a method described by Buntebarth (1978, 1979).

Samples, methods

Increased deep drilling activities of the Ruhr mining industry since 1974 have allowed coalification gradients to be determined at many places in the Ruhr Basin and the adjacent Lower Rhine mining district. The coalification studies were carried out by coal petrologists (M. Teichmüller and M. Wolf* of the Geological Survey of North Rhine-Westphalia at Krefeld during the years 1955-1979. Fig. 1 shows an example for the increase of coal rank with depth in the deepest borehole studied (1200 m), on the basis of different coalification parameters, all of which were determined

―――――――
* We gratefully acknowledge that Dr. M. Wolf placed her results at our disposal.

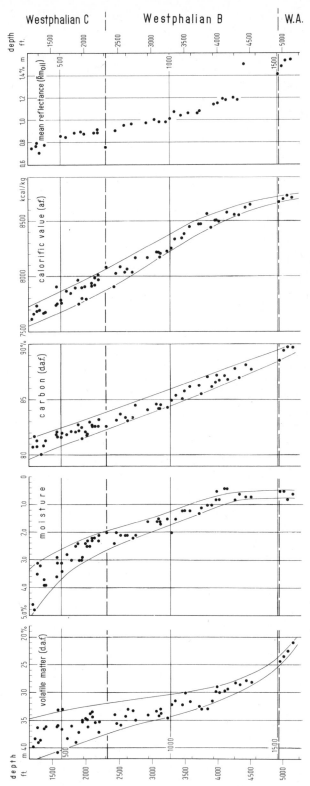

Fig. 1. The increase of degree of coalification with depth in the Nord-
licht Ost No.-1 borehole on the basis of vitrinite analysis.

on vitrite concentrates from coal beds. In this paper only the mean vitrinite reflectivity (% Rm) will be used as rank parameter. The optical reflectivity of vitrinite (at λ = 546 nm) was measured on polished sections of vitrite concentrates obtained from the different coal seams encountered in boreholes. Each Rm-value represents the arithmetical mean of 50 - 100 single measurements on the same polished block. The method of measuring vitrinite reflectivity is described in detail, inter alia, by Stach et al. (1975). 1591 reflectivity data obtained from seams in vertical sections of 53 boreholes were evaluated. The thickness of the Carboniferous in the bore profiles varies between 300 and 1200 m and is most frequently above 500 m. Stratigraphically the coal samples comprise Westphalian A, B and C. The rank ranges between 0.63 and 2.30 % Rm, in most cases between 0.8 and 1.9 % Rm (high-volatile bituminous A to low-volatile bituminous coal).

The evaluation was carried out by means of a method described by Buntebarth (1978, 1979), according to which the palaeothermal gradient is calculated on the basis of vitrinite Rm-values, the reconstructed maximal depth of burial and the duration of burial of the different coal seams.

The maximal burial is determined from the encountered thicknesses of the Westphalian A, Westphalian B (up to 820 m) and Westphalian C, considering the stratigraphic and geographical position of a given coal bed. Whereas the total thicknesses of Westphalian A and B remain more or less constant throughout the Ruhr Basin, the thickness of the Westphalian C varies and increases rapidly from south to north, as demonstrated by the iso-lines of Fig. 1.

The duration of burial follows from the span between sedimentation and upfolding of a given seam, the beginning of folding lying between Westphalian D and early Stephanian times. This span of time is surprisingly short. For the Katharina seam at the boundary of Westphalian A/B, for instance, it amounts to 15 million years only, thus being comparable with the coalification time in the molasse of the northern Alps.

By applying the geothermal regime of the Bavarian Molasse trough and that of the Upper Rhine Graben to the coalification conditions in some boreholes of these areas, an empirical relationship which describes the dependence of the degree of coalification on temperature and time of exposure was developed. If a constant temperature gradient, dT/dz, within a vertical column is assumed, the temperature is a function of depth. The depth is given by the burial history, $z(t)$.

In the young basins of southern Germany the square of the mean reflectivity of vitrinite, Rm^2 is related to a function of the temperature gradient, $f(dT/dz)$ and to the burial history in the following way (Buntebarth 1978, 1979):

$$Rm^2 = f(dT/dz) \cdot \int_{0}^{t} z(t^*) \, dt^*. \tag{1}$$

The maximal depth of a given sample is the sum of the present depth, $\tilde{z}$, and the former overburden, z_e, which has been eroded; hence

$$Rm^2 = f(dT/dz) \cdot \int_{0}^{t} \left(\tilde{z}(t^*) + z_e \right) dt^*. \tag{2}$$

The integration over the whole time range is not performable, because of the unknown burial history of the eroded part. Therefore, the integrated value is calculable from the time t_1 to the time t by splitting the integral into parts comprising the time interval from $t^* = 0$ to $t^* = t_1$ and $t^* = t_1$ to $t^* = t$:

$$Rm^2 = f(dT/dz) \left[\int_0^{t_1} (\tilde{z}(t*) + z_e) dt* + \int_0^{t_1} (\tilde{z}(t*) + z_e) dt* \right] \tag{3}$$

The first integral is an additive constant, c:

$$\int_0^{t_1} z_e \, dt* = c \tag{4}$$

to the variable second integral; $\tilde{z}(t)$ is not defined for $t < t_1$. With the use of expression (4), eq. (3) becomes:

$$Rm^2 = f(dT/dz) \left[\int_{t_1}^{t} (\tilde{z}(t*) + z_e) \, dt* + c \right] \tag{5}$$

The value c is either computable from the depth-time data and the corresponding reflectivity values by applying statistics or found graphically by plotting Rm^2 versus the integrated value. The axis intercept defines c.

To reduce the three-dimensional problem of eq. (5) to two dimensions, it is calculated at the plane $Rm = 1.0$ %. This procedure makes eq. (5) much easier to solve. At $Rm = 1.0$ % the integral reaches a value of I which is defined as:

$$I = \int_{t_1}^{t} (\tilde{z}(t*) + z_e) \, dt* + c \bigg|_{Rm = 1.0} \tag{6}$$

The function $f(dT/dz)$ is calibrated with coalification and temperature data of young sedimentary basins of southern Germany. The calibration yields the applied inverse function, f^{-1} of:

$$dT/dz \, [^\circ C/km] = 98.7 - 14.6 \ln I \, [10^6 a \cdot km]. \tag{7}$$

Results and discussion

Tab. 1 shows the calculated geothermal gradients, together with the depth intervals, the stratigraphy, the rank range (% Rm), and the number of Rm-values evaluated, for each borehole. For boreholes which encountered a sufficient thickness of Westphalian A and B + C, the geothermal gradients were calculated separately.

The coalification gradients range between 0.20 and 0.14 % Rm/100 m, the lowest gradients being determined within the lowest rank ranges. According to a diagram in which Rm-values are plotted versus relative depth (cf. M. & R. Teichmüller 1979, p. 223, Fig. 5.7), the increase of vitrinite reflectivity with depth is more or less constant in the Westphalian B+C sequences (mean: 0.048 % Rm/100m). In the Westphalian A the coalification gradients are markedly higher (mean: 0.104 % Rm/100m).

The highest geothermal gradients were found in boreholes which encountered Westphalian A. Fig. 2 shows a histogram of the frequency distribution of calculated temperature gradients for the Westphalian A and the Westphalian B+C. The gradients are remarkably higher for Westphalian A sequences with a mean value of 79°C/km than for Westphalian B + C sequences with a mean value of 65°C/km. The highest gradients (up to

Tab. 1. Vertical profile length, stratigraphy, rank range with the number of Rm-values evaluated and calculated temperature gradient for each bore-hole studied.

Number of the borehole	Vertical profile length (m)	Stage	Coalification range Rm (%)	Number of Rm values evaluated	Temperature gradient (^{o}C/km)
1	256	cwB	1.08-1.17	15	58
2	305	cwB	0.85-1.03	22	58
3	510	cwB	0.84-1.08	37	59
4	174	cwA	0.96-1.18	4	59
	352	cwB	0.92-1.10	22	
5	243	cwB	0.82-1.06	20	62
	77	cwC	0.85-0.95	9	
6	356	cwB	0.93-1.11	21	63
7	356	cwB	1.00-1.12	12	64
	32	cwC	0.97	1	
8	452	cwA	0.90-1.10	25	64
	219	cwB	0.87-1.00	7	
9	524	cwB	0.73-1.14	41	65
	70	cwC	0.82-0.93	7	
10	34	cwA	0.99-1.13	6	65
	469	cwB	0.85-1.07	34	
11	486	cwB	0.84-1.14	40	65
	72	cwC	0.81-0.93	6	
12	440	cwB	0.86-1.10	46	65
	322	cwC	0.77-0.88	24	
13	584	cwB	0.91-1.21	21	67
	377	cwC	0.76-0.90	11	
14	402	cwB	0.86-1.06	25	67
	437	cwC	0.63-0.88	19	
15	372	cwA	0.95-1.29	21	67 67
	277	cwB	0.85-0.98	17	56
16	331	cwB	0.86-1.07	23	67
17	205	cwA	1.07-1.28	10	67
	244	cwB	0.97-1.05	4	
18	293	cwA	0.98-1.24	21	67
	266	cwB	0.92-0.99	9	
19	593	cwB	0.86-1.28	50	69
20	553	cwB	0.72-1.23	50	69
21	220	cwA	1.17-1.43	11	69
	303	cwB	1.02-1.16	13	
22	292	cwA	1.16-1.45	12	69
	254	cwB	1.11-1.14	5	
23	525	cwB	0.86-1.30	45	70
24	377	cwA	1.02-1.39	19	73 71
	245	cwB	0.93-1.03	11	70

Number of the borehole	Vertical profile length (m)	Stage	Coalification range Rm (%)	Number of Rm values evaluated	Temperature gradient (°C/km)	
25	334	cwB	0.91-1.22	23	72	
26	532	cwA	0.93-1.38	34	72	
	44	cwB	0.93	1		
27	713	cwA	0.89-1.38	31	72	
	125	cwB	0.91-0.96	4		
28	412	cwA	1.40-1.72	20	74	
29	450	cwA	1.04-1.36	19	75	
	73	cwB	0.98-0.99	4		
30	494	cwB	0.72-1.19	40	76	
31	745	cwA	0.94-1.70	29	77	
32	448	cwA	1.16-1.56	19	78	
33	545	cwA	1.29-1.80	33	78	
34	707	cwA	0.90-1.56	83	78	
35	383	cwA	1.28-1.80	26	79	
36	219	cwA	1.17-1.40	19	83	
37	798	cwA	1.30-2.22	12	87	
38	150	cwA	1.41-1.62	8	87	
39	593	cwA	1.63-2.30	17	88	
40	387	cwA	2.42-2.88	8	89	
41	312	cwA	2.51-2.88	6	90	
42	404	cwA	1.65-2.29	14	90	
43	562	cwA	1.57-2.34	11	91	
44	300	cwA	0.99-1.26	17	69	63
	220	cwB	0.94-1.03	12	52	
45	180	cwA	1.09-1.23	10	73	63
	500	cwB	0.93-1.12	34	57	
46	440	cwA	0.99-1.36	28	75	72
	290	cwB	0.85-1.02	17	58	
47	260	cwA	1.03-1.26	17	75	67
	240	cwB	0.98-1.07	14	61	
48	615	cwB	0.90-1.20	19	68	
	339	cwC	0.70-0.91	17		
49	139	cwA	1.03-1.16	9	65	
	399	cwB	0.92-1.06	26		
50	68	cwA	1.27-1.29	3	67	
	339	cwB	1.06-1.19	10		
51	181	cwA	1.16-1.32	5	67	
	365	cwB	0.89-1.11	29		
52	286	cwB	0.92-1.20	28	67	
53	78	cwB	0.73-0.85	6	67	
	277	cwC	0.60-0.81	22		

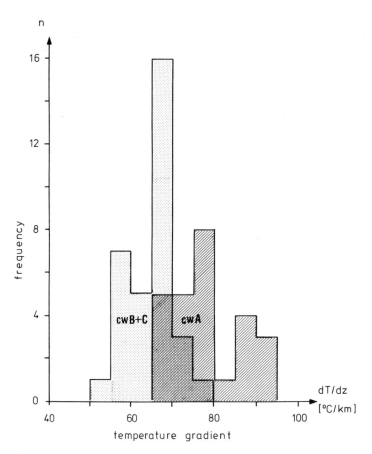

Fig. 2. Histogram of calculated thermal gradients in the Westphalian A and the Westphalian B+C.

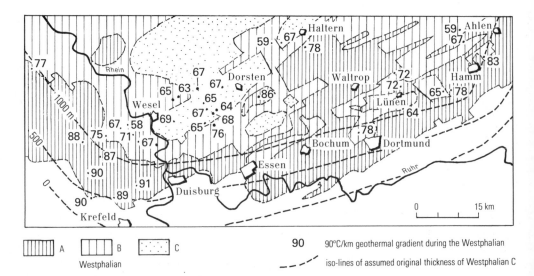

Fig. 3. The regional distribution of thermal gradients in the Carboniferous (Westphalian A to C) of the Ruhr and Lower Rhine Basins (sketch).

91°C/km) were calculated for the Westphalian A, north of the Krefeld high. These high values are due to a magmatic intrusion at great depth (Bunte-barth et al. 1981). Moreover, an anomalously high temperature gradient resulted from the coalification gradient found in the Westphalian A of the shaft N-3 at Westerholt, where the volatile matter yield of coal beds decreases rapidly with depth according to Wegehaupt (1962, p. 458, fig. 2). The rank parameter volatile matter can be easily converted into the rank parameter vitrinite reflectivity by using a relation between volatile matter of vitrites (d. a. f.) and mean vitrinite reflectivity (M. Teich-müller & R. Teichmüller 1979, p. 216, fig. 5.2). This allows using eq. (5) for the calculation of the geothermal gradient at Westerholt. From coalifi-cation gradients of 2.4 % volatile matter /100 m and 0.12 % Rm/100 m between the seams Albert 1 and Sonnenschein of the Westphalian A, a geo-thermal gradient of 86°C/km was calculated. This high gradient north of the central part of the Ruhr Basin differs substantially from the lower values of neighbouring boreholes (cf. Fig. 3). It is in agreement with the conclusion of Wegehaupt (1962) that an additional heat source was active in the Westerholt Block.

The lateral distribution of the palaeogeothermal gradients is de-monstrated in fig. 3, which also shows the occurrence of Westphalian A, B and C at the top Carboniferous in the Ruhr Basin. Furthermore, this sketchy map shows once again that the highest values correspond only to Westphalian A, i.e. at the western and eastern rim of the Ruhr Basin. However, the Westphalian A south of Haltern in the northern part of the basin is also characterized by a high thermal gradient compared with the gradients from neighbouring boreholes west of Haltern, which encountered Westphalian B and C. The high value (86°C/km) calculated for the Westerholt area south-east of Dorsten and the maximal values (89-91°C/km) found north of Krefeld have already been mentioned and explained above.

Some striking local variations of the calculated thermal gradients may have been caused by the different lengths of the depth intervals evaluated. The smaller the interval is, the more the temperature gradient is in-fluenced by local variations of the thermal conductivity. Moreover, the confidence in calculated temperature gradients depends on the number of Rm-values, which normally increases with the length of the depth interval.

The calculation of palaeotemperature gradients revealed a close re-lationship between the coalification gradients and the temperature gradients in the Ruhr Basin: the higher the coalification gradient, the higher the geothermal gradient. Moreover, there is a distinct relationship between the amount of the coalification and geothermal gradients and the level of coalification. The coalification level corresponding to more than 1.3 % Rm is characterized by a stronger increase of coal rank with depth than the coalification range between 0.65 and 1.3 % Rm (cf. M. & R. Teichmüller 1979, p. 223, fig. 5.7). This change of the coalification gradient is related to the "second coalification jump" at the stage of medium volatile bituminous coals, when a rapid release of hydrogen in the form of methane and a corresponding increase of aromatization begin.

It is striking that, in this study, the change of gradients coincides stratigraphically with the boundary between Westphalian A and Westphalian B. The higher coalification gradients (mean: 0.104 % Rm/100 m) and the resulting higher geothermal gradients (mean: 79°C/km) determined for the Westphalian A in contrast to the Westphalian B + C (mean values: 0.048 % Rm/100 m and 65°C/km respectively) may be explained if a higher thermal regime influenced the coalification of Westphalian A coal beds. Another explanation is a difference of coalification ranges encountered in the Westphalian A (mostly > 1.3 % Rm) and in the Westphalian B + C (mostly

< 1.3 % Rm), see above.

It is not possible to explain the different gradients by a different thermal conductivity, as a result of a different lithology of the various Westphalian stages for the following reasons: in the Westphalian A and B claystones predominate, whereas in the Westphalian C sandstones are abundant. Generally, sandstones have a higher thermal conductivity and may lead to lower coalification gradients (M. & R. Teichmüller 1968, p. 256, fig. 16). We thus should expect a change of gradients at the boundary between Westphalian B and C, which is not observed, however.

To estimate the heat flow density during the Westphalian, a thermal conductivity value must be applied. The low heat conductivity of coal, K = 0.26 W/m K, and the richness of coal within the coal-bearing rocks with K = 2 W/m K allow a mean conductivity of K = 1.6 W/m K to be assumed, if the clastic rocks contain 4 % of coal and if the heat flows in a perpendicular direction through the strata. If this conductivity value and the derived mean temperature gradients are employed, the heat flow during early Westphalian times was about $Q = 125 \text{ mW/m}^2$. This heat flow decreased to $Q = 105 \text{ mW/m}^2$ during the later Westphalian, as indicated by the rank of Westphalian B + C coals.

The apparent inverse relation between heat flow density and crustal thickness presented by Čermák (1979) allows a rough estimation of the crustal thickness during the Westphalian. By applying this relationship to the heat flow values estimated above, crustal thicknesses of about 20 km are evaluated for the early Westphalian, and of about 23 km for the later Westphalian. The recent crustal thickness in the Ruhr district is about 28 km (Prodehl et al. 1979). Of course, these estimated thicknesses of the crust must be considered cautiously. The assumed decrease of heat flow density to a lower but still high value during the Westphalian indicates an attenuating thermal regime at greater depth, which probably corresponds to a thickening of the crust according to empirical experiences (Čermák 1979).

There is no doubt that the marked difference between degree of coalification and geothermics in the northern foredeep of the Alps on one hand, and the Subvariscan foredeep in the Ruhr Basin on the other, must be attributed to different thicknesses of the crust (see introduction). The crust of the S1 Alpine foredeep is especially thick (30-40 km); this is probably due to the subsidence of the upper mantle within this subduction zone.

The high thermal gradients in the Subvariscan foredeep of the Ruhr area during the Westphalian correspond with the age of metamorphism of the Rhenish Variscan mountains. The K/Ar-method dates the peak of metamorphism in the northeastern Rhenish Variscan mountains at about 300 Ma before the present (Ahrendt et al. 1978), i. e. in the Westphalian. This coincidence emphasizes a heat supply from greater depth effective in both regions, the northeastern Rhenish Variscan mountains and the Subvariscan foredeep at the Ruhr.

References

Ahrendt, H., Hunziker, J. C. & Weber, K., 1978: K/Ar-Altersbestimmungen an schwach-metamorphen Gesteinen des Rheinischen Schiefergebirges. - Z. dt. geolog. Ges. 129: 229-247.

Buntebarth, G., 1978: The degree of metamorphism of organic matter in sedimentary rocks as a paleo-geothermometer, applied to the Upper Rhine Graben. - Pageoph. 117: 83-91.

Buntebarth, G., 1979: Eine empirische Methode zur Berechnung von paläo-
geothermischen Gradienten aus dem Inkohlungsgrad organischer Einlage-
rungen in Sedimentgesteinen mit Anwendung auf den mittleren Oberrhein-
Graben. - Fortschr. Geol. Rheinld. u. Westf. 29: 97-108.
Buntebarth, G., Michel, W. & Teichmüller, R., 1981: Das permokarbonische
Intrusiv von Krefeld und seine Einwirkung auf die Karbon-Kohlen am
linken Niederrhein. - Fortschr. Geol. Rheinld. u. Westf. 30: in press.
Čermák, V., 1979: Heat flow map of Europe. - In: V. Čermák & L. Rybach
(eds.): Terrestrial heat flow in Europe. - Berlin, Heidelberg, New York
(Springer), 3-40.
Giese, P. & Stein, A., 1971: Versuch einer einheitlichen Auswertung tiefen-
seismischer Daten aus dem Bereich zwischen Nordsee und Alpen. - Z.
Geophys., 37: 237-272.
Hallesches Jahrbuch für Geowissenschaften, Martin Luther-Universität,
Halle - Wittenberg.
Patteisky, K., Teichmüller, M. & Teichmüller, R., 1962: Das Inkohlungsbild
des Steinkohlengebirges an Rhein und Ruhr, dargestellt im Niveau von
Flöz Sonnenschein. - Fortschr. Geol. Rheinld. u. Westf. 3, II: 687-700.
Prodehl, C., Ansorge, J., Edel, J. B., Emter, D., Fuchs, K., Müller, St. &
Peterschmitt, E., 1976: Explosion-seismology research in the central
and southern Rhine Graben - a case history. - In: P. Giese, C. Prodehl
& A. Stein (eds.): Explosion seismology in central Europe. - Berlin,
Heidelberg, New York (Springer), 313-328.
Scheffer, V., 1965: The relation between the structure of the earth crust
and the hyperthermal territories. - Geofisika e Meteorologica 14: 49-
53.
Stach, E., Mackowsky, M. Th., Teichmüller, M., Taylor, G. H., Chandra, D.
& Teichmüller, R., 1975: Stach's textbook of coal petrology. - 2nd ed.,
Berlin, Stuttgart (Borntraeger), XII, 428 pp.
Teichmüller, M. & Teichmüller, R., 1968: Geological aspects of coal meta-
morphism. - In: Murchison, D. G. & Westoll, T. S.: Coal and coal-
bearing strata, 233-267.
- - , 1971: Die Inkohlung der Flöze im Rhein-Ruhr-Revier (Ruhrkohlen-
becken). - Fortschr. Geol. Rheinl. u. Westf. 19: 47-56.
- - , 1975: Inkohlungsuntersuchungen in der Molasse des Alpenvorlandes. -
Geologica Bavarica 73: 123-142.
- - , 1979: Diagenesis of coal, coalification. - In: Larsen, G. &
Chilingar, G. V. (eds.): Diagenesis in sediments and sedimentary rocks.
- Amsterdam, Oxford, New York (Elsevier), 207-246.
Teichmüller, R., 1973: Die paläogeographisch-fazielle und tektonische
Entwicklung eines Kohlenbeckens am Beispiel des Ruhrkarbons. - Z.
Deutsch. Geol. Ges. 124: 149-165.
Wegehaupt, H., 1962: Zur Petrographie und Geochemie des höheren Westfal A
von Westerholt. - Fortschr. Geol. Rheinld. u. Westf. 3: 445-495.

Heat flow map of North Atlantic

A. K. Popova, A. I. Ioffe and Ya. B. Smirnov

with 4 figures

Popova, A. K., Ioffe, A. I. & Smirnov, Ya. B., 1982: Heat flow map of North Atlantic. - Geothermics and geothermal energy, eds. V. Čermák & R. Haenel, E. Schweizerbart'sche Verlagsbuchhandlung, Stuttgart: 57-62.

Abstract: The heat flow map of the North Atlantic was constructed by two independent methods. The axis zone of the Mid-Atlantic ridge can be distinguished as an independent energetic system 120-250 km in width. The characteristic heat flow in the active rift zone of width 20-40 km is 200-400 mW m^{-2}.

Authors' addresses: A. K. Popova and A. I. Ioffe, Institute of the Physics of the Earth, USSR Academy of Sciences, Bolshaya Gruzinskaya 10, 123 242 Moscow, USSR; Ya. B. Smirnov, Geological Institute, USSR Academy of Sciences, Pyzhevsky per., 109 017 Moscow, USSR.

The construction of the heat flow map of the North Atlantic is based on a total of about 800 data (Foster et al. 1974, Popova 1974, Hyndman et al. 1976, Jessop et al. 1976, Savostin & Langseth 1977, Herman et al. 1977, Gorodnitskii et al. 1978, Poljak et al. 1978). In fact we constructed two independent heat flow maps (Fig. 1). The first map was constructed with the aid of a functional representation of the heat flow, and the other was created by simple interpolation of all tectonic, geophysical and other data.

   In using the functional representation we assume that the heat flow field q is described by the equation

$$q\ (\Phi,\delta) = \sum_{i=0}^{m} \sum_{j=0}^{n} a_{ij}\ \cos\ (j\ \Phi) P_i^j (\cos\ \delta) \tag{1}$$

where q is the heat flow at the point with coordinates Φ (latitude) and δ (longitude), P_i^j being the associated Legendre polynomial. To estimate the coefficients a_{ij}, we use the least squares method. In our case the value of m is 10; hence so we solved 66 normal equations and then constructed isolines by using the 60 x 60 grid of heat flow values.

   The zone of normal heat flow values (40-60 mW m^{-2}) occupies about 75 % of the area of the North Atlantic. With respect to its structure this zone coincides with the Atlantic basins and with the external flanks of the Mid-Atlantic ridge. On the background of the uniform heat flow field we can distinguish relatively small areas of anomalies of high (60-80 mW m^{-2}) heat flow and low (20-40 mW m^{-2}) heat flow.

   The corresponding heat flow distribution for the direction normal to the ancient fracture axis is shown in Figs. 1 B and 2. In the western province of the Atlantic Ocean, heat flow anomalies correspond to the Bagam fracture zone, the faults of the Bermuda uplift and the mountainous zone of New England. In the eastern province heat flow anomalies correspond to the

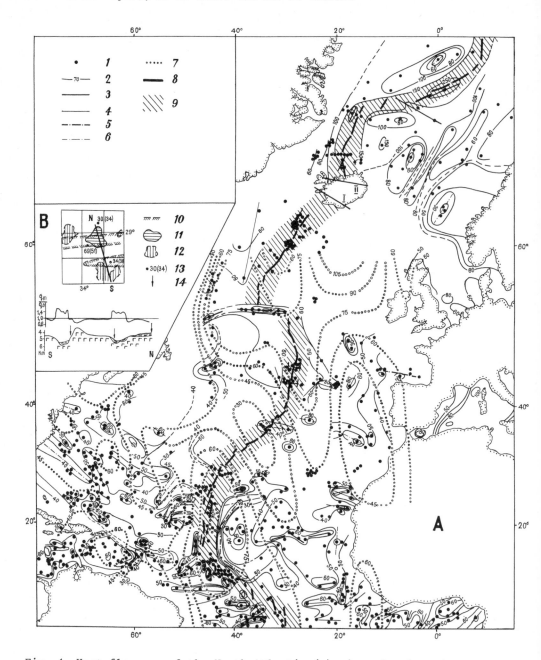

Fig. 1. Heat flow map of the North Atlantic (a); introduction of structure
correction along the Atlantic fault (b).
1 point of measurements; 2 isoline of heat flow in mW·m^{-2}; axis of geothermal anomalies (3 q > 100mW·m^{-2}; 4 60 < q < 100mW·m^{-2}; 5 q < 20 mW·m^{-2};
6 20 < q < 40 mW·m^{-2}); 7 isoline of functional representation of the heat
flow in mW·m^{-2}; 8 axis of Mid-Atlantic Ridge; 9 area of great variations
in heat flow values; 10 boundary of depression in the ocean bottom; 11
positive magnetic anomaly (> 100 γ); 12 negative magnetic anomaly (< −
100 γ); 13 values of measured (including correction) heat flow in mW·m^{-2};
14 points of measurements along the profile across the Atlantic fracture
zone.

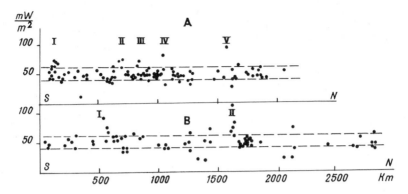

Fig. 2. Heat flow distribution of the directions normal to the transform faults axis in the North Atlantic.
A. The western part of the area: I Bagam fracture zone; II, III, IV Bermuda uplift faults; V mountains of New England.
B. The eastern part of the area: I the region of Cape Verde Islands, II Atlantic fault.

extension of the Atlantic fault to the eastern flank of the Mid-Atlantic ridge and to the zone near the Cape Verde Islands.

Geothermal anomalies in fracture zones are probably caused not by the deep processes, but rather by the distortion of the heat flow field by the structure of sediments and by the variations of the depth of the basement rocks.

The heat flow field in the adjacent Palaeozoic and Precambrian tectonic continental zones has the same structure often retaining the directions of anomalies.

Zones of high heat flow are distinguished in the Mid-Atlantic ridge extension to the north of Iceland, in the Norwegian Sea, in the rift parts of the transform faults, and in the internal zones of the Antilles Island arc.

Zones of low heat flow (~20 mW m^{-2}) limit the central part of the Mid-Atlantic ridge between 0°N and 30°N (there are few experimental data to the north of 30°N).

The dependence of heat flow on time for the time range of 0-10 Ma obtained with the data from three different parts of the North Atlantic is shown in Fig. 3. A minimum of the heat flow approaches the axis of the ridge in moving from the south to the north. The zone of the anomalously low heat flow near the axis of the ridge is characteristic of all oceans (Fig. 3A, Williams 1976). Hence it may be concluded that in the Atlantic, as well as in the other oceans, the minimum of the heat flow does not exceed the age interval of 3-10 Ma.

From the heat flow distribution in the central part of the ridge system the following features of the heat flow field in this region may be noted:
- The axis zones of the ridges in the range 0-6 Ma are the regions of very strong variations of heat flow.
- The narrow strips of anomalously low heat flow between magnetic anomalies N3 and N5 (about 6-10 Ma) are the specific feature of the heat distribution in the rift zone of the Mid-Atlantic ridge.
- In spite of such a great difference between the anomalies mentioned above, they all belong to the same tectonic structure – the rift zone of the Mid-Atlantic ridge.

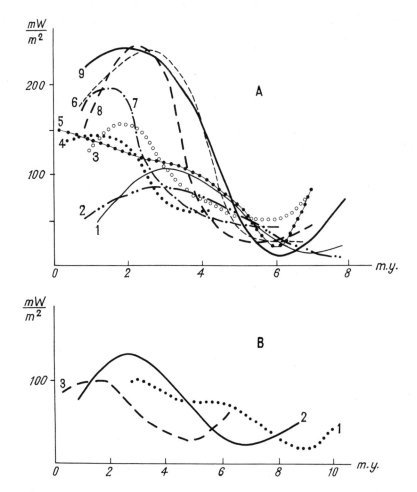

Fig. 3. Heat flow distribution in the axial parts of the rift ridges of world oceans (Williams 1976) and of the North Atlantic (B).
A. 1 Galapagos spreading centre at 86°W; 2 Mid-Indian ridge between 20°-22°S; 3 East Pacific rise at 14°S; 4 Juan-de-Fuca ridge; 5 East-Pacific rise at 12°S; 6 East Pacific rise at 39°S; 7 Reykjanes ridge; 8 Mid-Atlantic ridge at 46°N; 9 Gakke ridge.
B. 1 Vema fracture zone; 2 between the investigated fracture zone and the Azor zone; 3 between the Azor zone and the Gibbs fracture zone.

The mean heat flow value in the oceanic basins usually varies from 44 to 57 mW m^{-2}; however, the Labrador and Norwegian basins differ significantly from this mean, having higher heat flow values (over 60 mW m^{-2}). Some explanation can be sought in the presence of the ancient rifts in these basins and in the fact that the oceanic floors are significantly younger here. The uplifted regions of the ocean floor are generally characterised by the increased geothermal activity, as compared with the adjacent basins.

The heat flow fields in the oceanic basins and uplifts may be considered to be steady-state phenomena with a dominating conductive heat transfer component. The variations of the surface heat flow are rather small here and are caused by the local structural effects.

The mean value of heat flow is about 90 mW m^{-2} in the central part of the Mid-Atlantic ridge south of Iceland and about 195 mW m^{-2} north of Iceland for the range 0-6 Ma. The second time interval (6-10 Ma) is characterised by a mean value of heat flow ($\approx$ 50 mW m^{-2}) to the south of Iceland and 130 mW m^{-2} to the north of Iceland.

Thus the high background of the heat flow and the high standard deviation of the corresponding data are inherent in the recent active tectonic zones of the Atlantic. The regional variations of the heat flow in the rift zone of the Mid-Atlantic ridge (0-460 mW m^{-2}) cannot be explained by the distortion of the steady-state field near the surface, but an extensive convection must be assumed to exist below this zone.

The heat flow in transform faults in the North Atlantic is characterised by an anomalously high mean value which is fairly stable. In these zones the conductive heat transfer towards the surface is predominant because of the thick sedimentary layer. Hence the heat flow measured near the intersections of rift valleys and faults fits the value of the characteristic heat flow in the axial part of the rifts in the North Atlantic. The measurements for different parts of the Atlantic ridge gave values of heat flow in the interval of 200-400 mW m^{-2}, which also include the correction for sedimentation.

The analysis of the heat flow data in the Iceland region shows that the convective component of the total heat flow in the axis zone of the Iceland

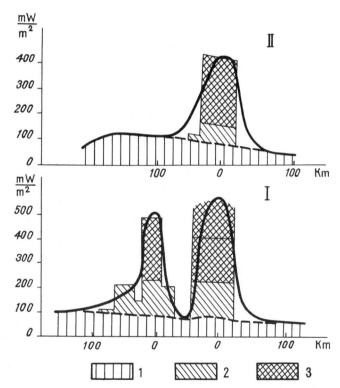

Fig. 4. Energetic balance for ancient and recent rifts of Iceland (Kononow & Poljak 1977, Poljak et al. 1978).
I the south profile; II the north profile; 1. Conductive heat transfer; 2. Hydrothermal effect; 3. Volcanism effect.

rift is 300–400 mW m^{-2} (Kononov & Poljak 1977) (Fig. 4). Since the energetic balance of the Iceland rift is probably the same as that of the rift of the Mid-Atlantic ridge, the convective part of the heat flow in the axial parts of the rift in the North Atlantic is 100–300 mW m^{-2}, and the conductive part is 100 mW m^{-2}.

Thus the analysis of geothermal data in the North Atlantic permits a distinction of the axis zone of the Mid-Atlantic ridge as an independent energetic system (width – 120–250 km). The width of the convective transfer zone (i.e. the active rift zone) is 20–40 km. The characteristic heat flow in the active rift zone is about 200–400 mW m^{-2}, of which roughly 25 % is due to the conductive and 75 % to the convective components.

References

Foster, S., Simmons, G. & Lamb, W., 1974: Heat flow near a North Atlantic Fracture zone. – Geothermics 3, N 1.

Gorodnitsky, A.M., Litvinov, E.M., Lukjanov, S.V. & Khutorskoy, M.D., 1978: Geomagnetic studies of the faults in the Atlantic and the structure of the heat flow field in this region. – In: Geophysical Methods of Investigation in the Arctic, 7–15. (Russ.)

Herman, B., Langseth, M.G. & Hobart, M.A., 1977: Heat flow in the oceanic crust bounding western Africa. – Tectonophysics 41: 61–72.

Hyndman, R.D. & Rankin, D., 1972: The Mid-Atlantic Ridge near 45°N. XVIII. Heat flow measurements. – Canad. Journ. Earth Sci., N 9: 664–670.

Jessop, A.M., Hobart, M.A. & Scater, J.G., 1976: The world heat flow data collection. – Geothermal series, Ottawa, N 5.

Kononow, B.I. & Poljak, B.G., 1977: Geothermal activity. – In: Iceland and Mid-ocean ridge. Deep structure, geothermics, 8–82. (Russ.)

Langseth, M.G. & Hobart, M.A., 1976: Interpretation of heat flow measurements in the Vema fracture zone. – Geoph. Res. Let. 3, N 5.

Poljak, B.G., Smirnov, J.B., Merkushov, V., Paduchikh, V.I. & Podgoenich, L.A., 1978: The new data about heat flow in area ridge Kolbensey. – Dokl. Acad. Sci. AN USSR, 243, N 1. (Russ.)

Popova, A.K., 1974: Geothermal measurements in oceans. – In: Geothermics. Reports about geothermal investigation in the USSR, N 1, 31–86. (Russ.)

Savostin, L.A. & Langseth, M.G., 1977: Geothermal investigation. – In: Iceland and Mid-ocean ridge. Structure of ocean bottom, 112–118. (Russ.)

Williams, D., 1976: Submarine geothermal resources. – Journ. Volcan. Geotherm. Res., N 1: 85–100.

Heat flow and heat generation in the new Gotthard tunnel, Swiss Alps (preliminary results) *

L. Rybach, Ph. Bodmer, R. Weber and Ph. C. England

with 3 figures

Rybach, L., Bodmer, Ph., Weber, R. & England, Ph. C., 1982: Heat flow and heat generation in the new Gotthard tunnel, Swiss Alps (preliminary results). - Geothermics and geothermal energy, eds. V. Čermák & R. Haenel, E. Schweizerbart'sche Verlagsbuchhandlung, Stuttgart: 63-69.

Abstract: The new Gotthard highway tunnel cross-sects over its length of 16.3 km the southern part of the Aar massif and the entire Gotthard massif; maximum cover: 1.5 km. During construction a substantial data base has been acquired: rock temperature measurements every 50 m, radioactivity measurements every 10 m, sampling for thermal conductivity and heat generation determinations every 100-200 m. Thermal conductivities vary between 2.1 and 4.0 W/m K, the anisotropy factor, in accordance with the macroscopic texture of the rock units traversed, from 1.0 to 2.3. Heat generation varies from 0.05 to 5.0 $\mu W/m^3$; a reasonable estimate can already be made on the basis of portable scintillation counter measurements in the tunnel. Subsurface temperatures have been corrected for the effects of topographic relief and of its evolution and of past climatic changes. The preliminary heat flow in the new Gotthard tunnel amounts to 52 mW/m^2 and supports independent evidence for subduction tectonism in the Swiss Central Alps during the alpine orogeny.

Authors' addresses: L. Rybach, Ph. Bodmer, R. Weber, Institute of Geophysics, ETH-Hoenggerberg, CH-8093 Zurich, Switzerland; Ph. C. England, Department of Earth Sciences, Bullard Laboratories, Madingley Rise, Cambridge, CB3 OEZ, England**

1. Introduction, geology

The new Gotthard tunnel, with its length of 16.3 km the longest highway tunnel of the world, offered a unique opportunity during its construction (1970-80) to sample unweathered rocks and to perform measurements of rock temperature and radioactivity. Radioactivity measurements (a total of over 1500) were made every 10 m along the tunnel axis, temperature measurements (over 400) were performed every 50 m and 120 representative rock samples were taken (every 100-200 m) for thermal conductivity and heat generation determinations.

The tunnel, located at a central position in the Swiss Alps (for geographical location see insert in Fig. 3), cross-sects the southern part of the Aar massif and the entire Gotthard massif; the maximum cover thickness is 1.5 km. The main lithologic units traversed are (from north to south):

* Contribution no. 324, Institute of Geophysics ETH Zurich
** Now at Department of Geological Sciences, Hoffmann Laboratory, Harward University, Cambridge, Mass. 02138 USA

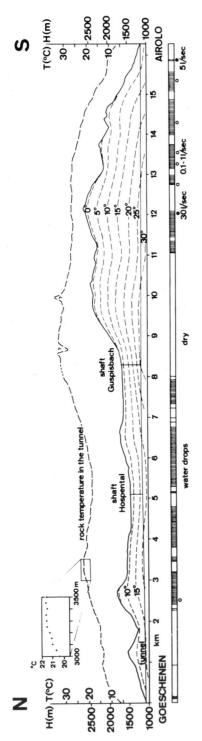

Fig. 1. Measured rock temperatures (top), topographic profile with constructed isotherms and hydrologic conditions (bottom) in the New Gotthard tunnel.

Central Aare granite, Southern gneissic cover of the Aar massif; Mesozoic and Permocarboniferous units, Paragneiss zone, Gamsboden granite, Guspis zone (mainly schists and amphibolites), Fibbia granite, Sorescia gneiss and Tremola series (metamorphic schists) of the Gotthard massif. Nearly all units show macroscopic signs of anisotropy (parallel texture); the planes of schistosity are in general vertical or near-vertical. A detailed petrographic description of the rock units traversed (including modal analyses) can be found in Rybach & Hafner (1962).

2. Subsurface temperatures, hydrology

During tunnel construction, *rock temperatures* were measured by applying a procedure developed by Dr. G. Berset (Zentralschweizerisches Technikum, Luzern) who also designed, built and periodically calibrated the thermistor instruments. The routine measurements were performed by the engineers of the constructing companies, at 50 m intervals, in 1.5-2 m deep, slightly downward inclined drillholes. These slim holes were specially drilled for the temperature measurements into the sidewall of the tunnel 1.5 m above floor. The temperature readings were taken 100-200 m behind the advancing front face within 5 to 10 days in order to avoid effects of ventilation.

The temperature readings were precise to $\pm$ 0.01 °C (G. Berset, personal communication). Cross-checks with measurements in the two deep vertical shafts of the Gotthard tunnel system (Hospental, 300 m; Guspisbach, 500 m) and against the temperature of the springs encountered in the tunnel (maximum thermometer readings) revealed that the rock temperature measurements can be considered as being reliable with an absolute accuracy of $\pm$ 0.2-0.3 °C.

The measured temperatures are plotted in Fig. 1; the enlarged portion illustrates the data density. The temperature curve follows roughly the course of the topography along the tunnel and reaches a maximum value of 31 °C at a distance of about 9 km from the north portal (Göschenen). The two peaks (one positive and the other negative) at 8.7 km and at 10.0 km could possibly arise from measurement error; at present we have no further explanation for them. The isotherms in Fig. 1 were constructed by interpolation between tunnel points (with the corresponding temperature) and the nearest surface point. The surface temperatures were determined from nearby meteorologic observations and the temperature measurements (at 15-20 m depth intervals) in the above-mentioned vertical shafts were also considered for the course of the isotherms.

With respect to the *hydrologic conditions* the new Gotthard tunnel can be considered, when compared to other deep tunnels in the Swiss Alps, as being relatively dry. E. g., extremely large water flows have been reported from the Simplon tunnel (up to 1000 l/sec, with a very low temperature of 10-14 °C, below a cover of 1170 m; Niethammer 1910). Comparatively little water was encountered in the new Gotthard tunnel. Zones of seepages and the entry points of small springs (the strongest issuing with 30 l/sec at 12 km distance from the north portal) are indicated in the bottom part of Fig. 1. Especially the northern part of the tunnel was dry. Therefore serious disturbances of the subsurface temperature field by circulating water can be ruled out.

3. Rock radioactivity, heat generation

Rock radioactivity measurements (gamma radiation) were performed in the tunnel by a portable scintillation counter (type SRAT). Readings were taken

in cps (counts per second; 100 cps ≈ 25 µr/h). Considerable variation of
the integral gamma counts was found along the tunnel (from ~0 cps to over
600 cps) but readings were fairly constant within the individual lithologic
units. In the Central Aare granite a systematic increase of rock radioactivity from north to south (from 200 to 300 cps) was found (see also
Labhart & Rybach 1980).

Radioelement contents and heat generation were determined in the
laboratory on the rock samples by gamma ray spectrometry (for details of
the technique applied see Rybach 1976a). Heat generation of individual
samples varied with rock type from 0.05 to 5.0 µW/m^3.

Reasonable estimates of heat generation can already be made from the
scintillation counter measurements in the tunnel, by means of the empirical
relationship depicted in Fig. 2. In this diagram, mean gamma activities
(in cps) of the individual lithologic units, as measured in the tunnel,
are plotted against the mean heat generation values of corresponding rock
samples. A clear linear relationship is evident; the reason for the strong
positive correlation (r = 0.977) can be seen in the relatively constant
measurement geometry in the tunnel (4Π solid angle) and also in fairly
constant Th/U and K/U ratios. Due to different ventilation conditions (and
thus different radon content in the tunnel air) the zero intercept of this
linear relationship can vary significantly from one tunnel to another.

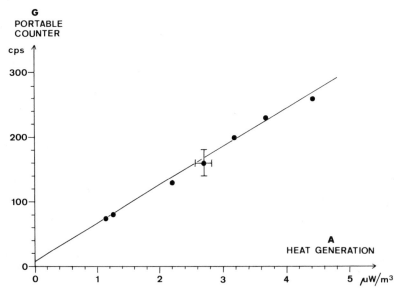

Fig. 2. Empirical relationship between rock radioactivity measurements in
the tunnel (scintillation counter readings in cps; 100 cps ≈ 25 µr/h) and
heat generation determinations on corresponding rock samples. Mean values
of the main lithologic units are plotted. Regression analysis yields
G (cps) = 14.3 + 57.4 A (µW/m^3) with r = 0.977.

4. Results, conclusions

For heat flow determinations the subsurface temperature measurements must
first be corrected to calculate the geothermal gradient. In the physio-
graphic region under study (Swiss Central Alps) corrections have to be
applied to account for the effects of rugged topography, recent uplift/

erosion and past climatic changes. Three-dimensional topographic corrections at the rock temperature measurement points in the tunnel, taking into
account the time of evolution of the topography as well, are based on an
array covering the area of whole Switzerland and digitized (topographic
heights) at mesh points every 250 m. The effect of erosion is also calculated and finally the geothermal gradient is determined by a least squares
fit. For a detailed description of the calculation procedure see Bodmer
et al. (1979). The uncorrected gradient for all temperature measurements
in the tunnel is 16.0 °C/km, after topographic correction 23.7 °C/km and
after erosional correction 19.0 °C/km (regression coefficient 0.87). The
palaeoclimatic effect on the gradient is estimated, by taking into account
the temperature change at the end of the last period of glaciation (+ 4 °C,
14 000 years ago), to be -1.9 °C/km for an average depth (cover thickness)
of 1 km.

Thermal conductivities were measured on the collected rock samples by
the line source method with a QTM (Quick Thermal Conductivity Meter, Showa
Denco Co.) equipment. Special attention was payed to the anisotropy effect.
For heat flow calculation the mean thermal conductivity in the vertical
direction was determined for each lithologic unit and weighted according
to the thickness of these zones. The weighted average vertical thermal
conductivity of the rocks in the new Gotthard tunnel is 3.02 W/m K and the
weighted heat generation is 2.98 $\mu W/m^3$.

The preliminary heat flow value for the new Gotthard tunnel, 52 mW/m^2,
is plotted against the weighted mean heat generation in Fig. 3. The Gotthard tunnel heat flow point falls very close to the line which represents
the Sierra Nevada continental heat flow province (see Roy et al. 1968).
The low reduced heat flow (zero intercept of the straight line) for the

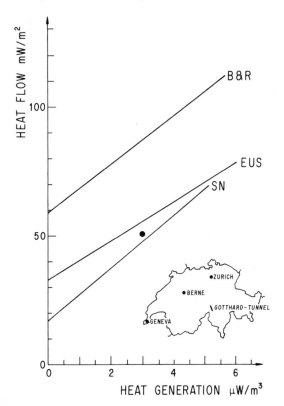

Fig. 3. Heat flow - heat
generation plot. The value for
the new Gotthard tunnel falls
close to the line that represents the Sierra Nevada-type
continental heat flow province
(SN). The other lines indicate
further heat flow provinces
(B & R: Basin and Range USA,
EUS: Eastern United States)
according to Roy et al. (1968).

Sierra Nevada province has been interpreted to be indicative of the subduction tectonism which affected the Sierra Nevada mountain range during
early to mid-Tertiary (transient effect of the cold subducted slab, see
Lachenbruch & Sass 1977).

For the Central Alps subduction tectonism during the alpine orogeny
has been suggested by several authors (see e. g. Laubscher 1977) and
corresponding thermal effects were postulated by Rybach (1976b). Impressive
seismic evidence for the presence of subducted lithospheric remnants below
the Central Alps was presented by Panza & Mueller (1979). The above-
mentioned findings (low heat flow, low reduced heat flow in the Gotthard
area) support the idea of subduction. The heat flow value reported for the
old Gotthard railroad tunnel (67 mW/m^2, Clark & Niblett 1956) is low as
well but was not corrected for the effect of erosion (which would lower
the value by 20 %). The same applies to the mean value of heat flow
determinations in road and hydroelectric tunnels in the Tauern area,
Eastern Alps (Oxburgh & England 1980).

Acknowledgement. The rock temperature data were supplied by Dr. T. R.
Schneider, Prof. E. Dal Vesco and Dr. H. Wanner to whom we also owe special
thanks for many stimulating discussions about the geology of the new
Gotthard tunnel. Thanks are also due to Ueli Schärli and Werner Rellstab
for the thermal conductivity measurements. This work has been supported
by the Swiss National Science Foundation, Projects no. 2.616.-0.76 and
2.108.-078.

References

Bodmer, Ph., England, P.C., Kissling, E. & Rybach, L., 1979: On the
    correction of subsurface temperature measurements for the effects of
    topographic relief, Part II: Application to temperature measurements in
    the Central Alps. - In: V. Čermák & L. Rybach (Eds.), Terrestrial Heat
    Flow in Europe. - Springer-Verlag, Berlin, Heidelberg, New York, pp.
    78-87.
Clark, S.P. & Niblett, E.R., 1956: Terrestrial heat flow in the Swiss Alps.
    - Royal Astron. Soc. Geophys. Suppl. Monthly Notices 7: 176-195.
Labhart, T.P. & Rybach, L., 1980: Radiometrische Stollenprofile durch
    zentralmassivische Granite im Bereich der Schweizer Geotraverse. -
    Eclogae geol. Helv. 73: 571-582.
Lachenbruch, A.H. & Sass, J.H., 1977: Heat flow in the United States and
    thermal regime of the crust. - In: J.G. Heacock (Ed.), The Earth's
    Crust - Its Nature and Physical Properties. - Amer. Geophys. Union
    Monograph no. 20, pp. 626-675.
Laubscher, H.-P., 1977: The tectonics of subduction in the Alpine system.
    - Mem. Soc. geol. ital. Suppl. 2, 13: 275-283.
Niethammer, G., 1910: Die Wärmeverteilung im Simplon. - Eclogae geol. Helv.
    11: 96-120.
Oxburgh, R.E. & England, P.C., 1980: Heat flow and metamorphic evolution
    of the Eastern Alps. - Eclogae geol. Helv. 73: 379-398.
Panza, G.F. & Mueller, St., 1979: The plate boundary between Eurasia and
    Africa in the Alpine area. - Mem. Sci. geol. 33: 43-50.
Roy, R.F., Blackwell, D.D. & Birch, F., 1968: Heat generation of plutonic
    rocks and continental heat flow provinces. - Earth Planet. Sci. Lett.
    5: 1-12.

Rybach, L., 1976a: Radioactive heat production in rocks and its relation to other petrophysical parameters. - Pure and Appl. Geophys. 114: 309-318.

— — , 1976b: Die Schweizer Geotraverse Basel-Chiasso: Eine Einführung. - Schweiz. Min. Petr. Mitt. 56: 581-588.

Rybach, L. & Hafner, St., 1962: Radioaktivitätsmessungen an Gesteinen des St. Gotthard-Profils. - Schweiz. Min. Petr. Mitt. 42: 209-219.

Heat flow density determination in shallow lakes along the geotraverse from München/Salzburg to Verona/Trieste

R. Haenel and G. Zoth

with 4 figures and 2 tables

Haenel, R. & Zoth, G., 1982: Heat flow density determination in shallow lakes along the geotraverse from München/Salzburg to Verona/ Trieste. - Geothermics and geothermal energy, eds. V. Čermák & R. Haenel, E. Schweizerbart'sche Verlagsbuchhandlung, Stuttgart: 71-78.

Abstract: Six new heat flow density values were determined in shallow lakes along the geotraverse and 26 existing values from deep lakes and boreholes were revised. The influence of topography, refraction caused by differing thermal conductivities of the rock, sedimentation, and paleoclimatic temperature effects have been eliminated. A simplified isoline map shows five regions: the Molasse Basin and Po Basin with about 80 mW m^{-2}, followed by zones towards the centre of the Alps with about 125 mW m^{-2}, and the centre of the Alps with about 65 mW m^{-2}.
The heat flow density pattern can be explained for example by the assumption of normal conductivity values and heat production in rock in the subsurface. The influence of uplift and denudation of the Alps is relatively small.

Authors' address: Geological Survey of Lower Saxony, P.B. 510 153, D-3000 Hannover, Federal Republic of Germany.

Introduction

In the past, heat flow density was determined in deep lakes (Haenel 1974). This was possible because the annual temperature change at the bottom of a deep lake is usually smaller than about 1 $^{\circ}$C, and its influence on the temperature gradient in the sediment could be easily eliminated. In shallow lakes, the amplitude of the temperature changes at the lake bottom can amount to several degrees. To correct for such influences, the temperature changes must be known for about the past year.

Measurements

In the summer 1975 temperature stations were set up at the bottom and in the middle of 15 lakes in Austria and Italy along a geotraverse between München/Salzburg and Verona/Trieste.

These temperature stations consist of a platinum element and a battery-driven recording unit within a pressure vessel 64 cm long and 16 cm in diameter. The accuracy is better than 0.2 $^{\circ}$C (Zoth 1978).

After one year, in 1976, the temperature stations were removed, and heat flow density measurements were carried out at the same place with a so-called lake probe (Haenel 1979). Only 6 stations could be recovered, one was blocked by landslide, one was pulled out by a fisherman, another one by children, another could'nt be found etc.

The results are compiled in Tab. 1 under numbers 1 to 6.
Corrections were made for the following:
- inclination of probe;
- the influence of annual temperature wave at the lake bottom on the
  temperature gradient in the sediment;
- topography, including the effect of the water body;
- the effect of the different thermal conductivities of the sediment and
  surrounding rock using estimated values for both, including known values
  from our own measurements.

Tab. 1a. Results of heat flow density values measured in shallow lakes.

No.	Name	Date	Co-ordinates		Water level
			(Grad-Min)	(Grad-Min)	
1	Walchsee	16.10.76	47 - 38.8	12 - 19.7	655
2	Hechtsee	15.10.76	47 - 36.6	12 - 9.9	542
3	L.d. Levico	25.10.76	46 - 1.5	11 - 16.2	440
4	L.d. Lases	24.10.76	46 - 8.0	11 - 15.3	632
5	L.d. Cavedine	24.10.76	46 - 1.0	10 - 57.0	241
6	L.d. Caldonazzo	27.10.76	46 - 0.7	11 - 15.0	450

No.	Water depth (m)	Mean thermal conductivity (W/m^oC)	Mean temperature gradient $(^oC/m)$	Mean heat flow density (mW/m^2)
1	21	0.750	0.0531	39.8
2	55	0.584	0.1336	79.6
3	36	0.875	0.2529	221.3
4	25	0.670	0.1682	112.7
5	47	0.712	0.0900	64.1
6	46	1.014	0.1371	139.0

Tab. 1b. Improvement of heat density values by corrections

No.	Name	Mean heat flow density (mW/m^2)	Combined Correction (%) 100%	(mW/m^2)
1	Walchsee	39.8	105	37.9
2	Hechtsee	79.6	117	68.0
3	L.d. Levico	221.3	175	126.5
4	L.d. Lases	112.3	207	54.2
5	L.d. Cavedine	64.1	120	53.3
6	L.d. Caldonazzo	139.0	112	124.1

No.	Paleoclimatic Correction + 0.0125 $^oC/m$	Correction of sedimentation (%)	Corrected mean heat flow density (mW/m^2)
1	46.9	+15	53.9
2	74.3	+15	85.4
3	132.7	+15	152.6
4	58.5	+15	67.3
5	60.8	+15	70.0
6	135.4	+15	155.7

In addition, the influences of sedimentation is assumed to be + 15 % and for lakes higher than 700 m above sea level + 3 % on the basis of the sedimentation rate of Alpine lakes determined by Finckh (1976).

The paleoclimatic influence is considered to be a mean values of 0.0125 $^{o}C\ m^{-1}$ ± 33 % for all values in this paper as shown by Finckh (1976).

These corrections, especially the last two, have also been applied to already existing lake values along the above-mentioned geotraverse; see Tab. 2, nos. 7 - 33.

The values are also shown on the map in Fig. 1. With regard to the high scatter, isolines were drawn separating them into values $> 100\ mW\ m^{-2}$ and

Fig. 1. The heat flow density ($mW\ m^{-2}$) distribution along the geotraverse München/Salzburg - Verona/Trieste. The shadow zone represents values with more than 100 $mW\ m^{-2}$.

Tab. 2. The already existing heat flow density values measured in lakes
and improved as described in the text

No.	Name	Hitherto existing values ($mW\ m^{-2}$)	Improved values ($mW\ m^{-2}$)
7	Ammersee	71.6	
8	Alpsee	87.9	125.9
9	Königsee	76.2	176.7
10	Schliersee	63.6	78.8
11	Starnberger See	84.6	88.4
12	Walchensee	75.8	101.0
13	Brg. Peißenberg	75.4	75.4
14	Achensee	88.4	119.4
15	Altauseer See	49.0	82.8
16	Attersee	89.6	130.5
17	Fuschl See	87.5	143.4
18	Hallstätter See	63.2	107.2
19	Millstätter See	73.3	112.0
20	Mondsee	52.3	67.5
21	Plansee	88.4	171.8
22	St. Wolfgang See	67.8	111.6
23	Toplitzsee	67.0	146.3
24	Traunsee	127.3	205.1
25	Weißensee	67.4	100.4
26	Zellersee	81.1	109.9
27	Arlberg Tunnel	89.6	89.6
28	Tauern Tunnel	80.4	80.4
29	Brg. Tauern	59.9	59.9
30	Brg. Bleiberg	47.0	47.0
31	Mittersill	37.7	37.7
32	Gardasee-Mitte (Haenel)	92.7	111.8
33	Gardasee-Mitte (Finckh)		118.7

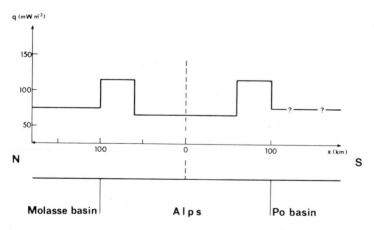

Fig. 2. The simplified heat flow density distribution derived from Fig. 1.

< 100 mW m^{-2}. The surprising result is an approximately heat flow density distribution as shown in Fig. 2.

Interpretation

The simplified profile in Fig. 2 is possibly caused by the following phenomena:
- In the Central Alps (region of low heat flow density), water is descending to great depth, and the heated water ascends along the edge of the Alps (region of high heat flow density). This would be in agreement with the experience that high heat flow density values are mostly connected with water upwelling from great depth, but not reaching the earth's surface. This hypothesis can not be proved because no information is available.
- The high values (edge of the Alps) and low values (Molasse basin and Po basin) could be explained by the subduction zone theory. But the presence of two subduction zones within such a small area has not been demonstrated anywhere. On the other hand, the heat flow density could be caused by a classical horizontal shifting of rock. An examination of that is out of the scope of this paper.
- The heat flow density pattern can be explained by differences in lateral thermal conductivity, as well as heat production by radioactive elements. This can be easily studied.

1. Steady-state models

First, the influences of different thermal conductivity at the transition Alps/foreland basin can be studied. This is done by using the equation:

$$\text{div } (\lambda \text{ grad } T) = 0.$$

This is solved using the following differential equation:

$$\lambda \left(\frac{\partial^2 T}{\partial x^2} + \frac{\partial^2 T}{\partial z^2} \right) + \left(\frac{\partial \lambda}{\partial x} \frac{\partial T}{\partial x} + \frac{\partial \lambda}{\partial z} \frac{\partial T}{\partial z} \right) = 0.$$

Assuming reasonable thermal conductivity values of $\lambda = 2.9$ (2.7) W m^{-1} $^{\circ}$C^{-1} for the Alps and $\lambda = 1.7$ (1.9) W m^{-1} $^{\circ}$C^{-1} for the foreland basins, this equation yields the results shown in Fig. 3. The calculation has been carried out for a flat area and for a two-level area which corresponds in principle to the topography of the Alps and the foreland basins.

2. Non-steady-state models

Considering the distribution of thermal conductivity and heat production, the uplift and denudation of the Alps, and the subsidence of the basins, the following equation fits the problem:

$$\text{div } (\lambda \text{ grad } T) + H = \rho c \frac{\partial T}{\partial t} + \rho c \vec{v} \text{ grad } T.$$

The equivalent differential equation is:

$$\lambda \left(\frac{\partial^2 T}{\partial x^2} + \frac{\partial^2 T}{\partial z^2} \right) + \left(\frac{\partial \lambda}{\partial x} \frac{\partial T}{\partial x} + \frac{\partial \lambda}{\partial z} \frac{\partial T}{\partial z} \right) + H = \rho c \frac{\partial T}{\partial t} + \rho c \, v_z \frac{\partial T}{\partial z} ,$$

where H = heat production, ρ = density, c = specific heat capacity, t = time, v_z = rate of uplift and denudation. The approximation method of Minear & Toksöz (1970) is used to solve this equation.

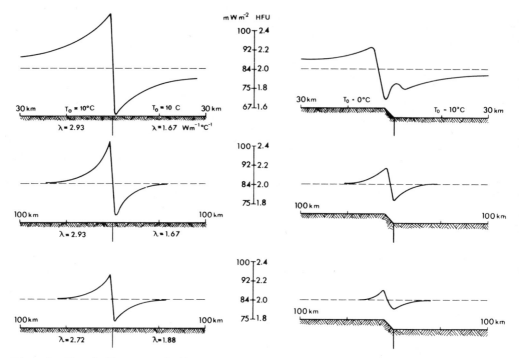

Fig. 3. The influence of different thermal conductivities on the heat flow density along the border Alps/basin. For the models on the right the same thermal conductivity values are used as on the left.

On the basis of data given by Angenheister et al. (1975), Bögel & Schmidt (1976), and Cliff et al. (1971), the following simplified assumptions have been made:
- Between 40 and 20 million years ago the foreland basin were created by continuous subsidence of 0.25 mm per year to 5 km depth.
- Between 20 million years ago and the present, the Alps have been up-lifted at a rate of 0.75 mm per year. Simultaneously the Alps are being denudated.

Two models have been calculated with the above-mentioned conditions:
Model 1: Initial value for heat flow density is 82 mW m^{-2}, for the thermal conductivity within the Alps 2.93 W m^{-1} $^{\circ}$C^{-1} and outside the Alps 1.67 W m^{-1} $^{\circ}$C^{-1}. Acid rock, metamorphic shist, gneiss and sediments, and a heat production of H_g = 1.46·10^{-6} W m^{-3} are assumed between 0 and 15 km depth (Rybach 1973, Haenel 1971). Basic rock and a heat production of H_b = 0.67·10^{-6} W m^{-3} are assumed between 15 km depth and the Mohorovicic dis-continuity. Basic material ascends in the Central Alps (60 km north to south, corresponding to the Tauern window) from a source at a depth of 20 km; acid material ascends on both sides of this area, but still within the Alps, from a 15 km deep source.
Model 2: Initial value for heat flow density is 78 mW m^{-2}, for thermal conductivity as in Model 1. H_b is assumed to be H_b = 2.92·10^{-6} W m^{-3}, and

the width of the area of basic material in the center is enlarged from 60 to 120 km.

Angenheister's (1972) value for the depth of the Mohorovicic discontinuity was used. A topographic step between the Alps and foreland basin has been neglected. In the computer model, a horizontal distance of 10 km has been assumed between points in the grid, a vertical distance of 1 km, and a time interval of 10,000 years.

Results

The simplified heat flow density distribution of Fig. 2 is repeated in the upper part of Fig. 4. The mean values lead to 5 zones:

Northern Molasse basin (78 mW m^{-2}), northern Alps (128 mW m^{-2}), central Alps (65 mW m^{-2}), southern Alps (126 mW m^{-2}), and southern Po basin (78 mW m^{-2}, assumed value).

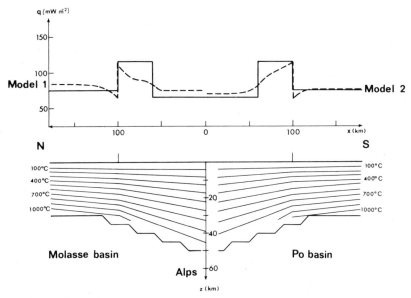

Fig. 4. Lower part: temperature distribution down to the Mohorovicic discontinuity for model 1 (left side) and model 2 (right side). Upper part: heat flow density distribution on the earth's surface from Fig. 1 (full line) and from the model calculation (dashed line).

The calculated heat flow density distribution, as well as the calculated temperature distribution of Model 1, is presented on the left side of Fig. 4 and from Model 2 on the right side.

There is a quite reasonable correspondence of the calculated heat flow density distribution of Model 1 and 2 with the simplified heat flow density distribution calculated from the collected measurements. A better correspondence can be expected by improved modeling.

The influence of uplift and denudation is not recognizable. From the intermediate computer results - which are not presented here - it can be seen that this influence is lower than by the calculation carried out by

Haenel & Zoth (1973). There is no influence on heat flow density values at present from the subsidence of the basin structures.

It can be seen that the temperature isolines calculated by Models 1 and 2 are more or less parallel to the Mohorovicic discontinuity.

Acknowledgement. The Deutsche Forschungsgemeinschaft, Bonn - Bad Godesberg, is thanked for the financial support which enabled these heat flow density measurements carried out. The authors also wish to express their thanks to Dipl.-Ing. P. Melichar, Zentralanstalt für Meteorologie und Geodynamik (Wien) and Dipl.-Geoph. H. Villinger, Technische Universität (Berlin) for their helpful support, and to Dr. Newcomb who kindly corrected the English text.

References

Angenheister, G., Bügel, H., Gebrande, H., Giese, P., Schmidt-Thomé, P. & Zeil, W., 1972: Recent investigations of surficial and deeper crustal structures of the Eastern and southern Alps. - Geol. Rdsch. 61, 2: 349-395.

Angenheister, G., Bügel, H. & Morteani, G., 1975: Die Ostalpen im Bereich einer Geotraverse vom Chiermsee bei Vicenza. - N. Jb. Geol. Paläont. Abh. 148, 1: 50-137.

Bügel, H. & Schmidt, K., 1976: Kleine Geologie der Ostalpen. Allgemein verständliche Einführung in den Bau der Ostalpen unter Berücksichtigung der angrenzenden Südalpen. - Ott-Verlag, Thun, S. 1-231.

Cliff, R.A., Norris, R.J., Oxburgh, E.R. & Wright, R.C., 1971: Structural, metamorphic and geochronical studies in the Reisseck and southern Ankogel Groups, the eastern Alps. - Jahrb. Geol. B.-A. Wien 114: 121-272.

Finckh, P., 1976: Wärmeflußmessungen in Randalpenseen. - Mitt. Geol. Inst. Eidg. Techn. Hochschule Univ. Zürich, Dissert. Nr. 5787: 1-105.

Haenel, R., 1971: Heat flow measurements and a first heat flow map of Germany. - Z. Geophys. 37: 975-992.

— , 1974: Heat flow measurements in northern Italy and heat flow maps of Europe. - J. Geophys. 40: 367-380.

— , 1979: A critical review of heat flow measurements in sea and lake bottom sediments. - In: Terrestrial Heat Flow in Europe, ed. by V. Čermák & L. Rybach, Springer Verlag, 49-73.

Haenel, R. & Zoth, G., 1973: Heat flow measurements in Austria and heat flow maps of Central Europe. - Z. Geophys. 39: 425-439.

Haenel, R., Grönlie, G. & Heier, K.S., 1974: Terrestrial heat flow measurements from lakes in southern Norway. - Norsk Geologisk Tidsskrift, Oslo, 54, 4: 423-428.

Minear, J.W. & Toksöz, M.N., 1970: Thermal regime of a downgoing slab and new global tectonics. - J. Geophys. Res. 75: 1397-1419.

Rybach, L., 1973: Wärmeproduktionsbestimmungen an Gesteinen der Schweizer Alpen. - Beitr. Geol. Schweiz. Geotech. S., L. 51, 1-43, Zürich (Leemann AG).

Zoth, G., 1978: Temperaturmessungen in einigen norwegischen Seen. - Report, NLfB-Archives, No. 79 077, Hannover.

Some aspects of heat flow in France

Guy Vasseur and Francis Lucazeau (Groupe Fluxchaf)

with 4 figures and 3 tables

Vasseur, G. & Lucazeau, F., 1982: Some aspects of heat flow in France. - Geothermics and geothermal energy, eds. V. Čermák & R. Haenel, E. Schweizerbart'sche Verlagsbuchhandlung, Stuttgart: 79-89.

Abstract: The basic information about heat flow in France consists of several types of data which can be separated according to their measurement characteristics. After consideration of the necessary corrections (topographic and palaeoclimatic effects) a critical comparison between the various types of data is carried out. This comparison brings out a clear discrepancy between some of the data: in the Paris basin, heat flow estimates resulting from oil exploration boreholes without conductivity measurements are systematically higher than other types of measurements. Keeping only the more reliable heat flow measurements, we present a tentative heat flow map which accounts for the main trend of the remaining data. The most significant feature of this map is the presence of a high heat flow area in the central part of the country; this positive anomaly seems to be related to the existence of Hercynian granites and, possibly of crustal and lithospheric thinning.

Authors' address: Centre Géologique et Géophysique, Université des Sciences et Techniques du Languedoc, Place E. Bataillon, F-34060 Montpellier, France.

Introduction

In recent years and following the first heat flow measurements published by Hentinger & Jolivet (1970), a campaign has been developed by the Bureau de Recherches Géologiques et Minières and by several university teams sponsored by the Institut National d'Astronomie et de Géophysique (Groupe Fluxchaf) for the determination of heat flow in France. The main objective was to obtain reliable values of heat flow in various areas in order to construct an isoline map; several attempts in this direction leading to different models have already been proposed (Groupe Fluxchaf 1978, Gable & Goguel 1978, Vasseur & Nouri 1980). In fact a major difficulty results from the large differences in the characteristics of the various heat flow determinations. For example, some data are obtained by means of conductivity measurements in shallow boreholes, whereas others are estimations from deep exploration boreholes. Hence a preliminary discussion is necessary before such data can be used simultaneously.

In this paper, the available heat flow determinations are reviewed and discussed; relevant corrections for topography and palaeoclimatic effects are carried out and the consistency of the data is studied. On the basis of this discussion, a tentative heat flow map is then presented and briefly compared with geological features.

Tab. 1. List of data of group B.
NO: reference number - Station: name of station - LAT, LONG: Geographical coordinates - ALT: Ground altitude (in m) - N: number of boreholes used (N = 0: in mines) - ZMIN, ZMAX: Minimum and maximum depth used for measurements - NM: number of individual heat flow measurements - KM: mean value of thermal conductivity (W m^{-1} ^{o}C^{-1}) - FX1: uncorrected heat flow in mW m^{-2} - FX2: heat flow corrected for topographic effect - FX3: heat flow corrected for topographic effect and palaeoclimatic effect (post-glacial warming of 8 ^{o}C) - FX4: heat flow corrected for topographic effect and palaeoclimatic effect (postglacial warming of 14 ^{o}C).

NO	Station	LAT	LONG	ALT	N	ZMIN	ZMAX	NM	KM	FX1	FX2	FX3	FX4
091	Bac de Montmeyre	45.46	2.56E	945	1	100	240	15	2.3	73	70	77	84
092	Bournac	44.59	4.06E	1100	1	50	190	15	2.5	86	86	95	102
093	Mayet de Mont	46.04	3.38E	570	1	20	150	20	3.0	80	84	93	102
094	Montalba	42.40	2.33E	480	1	30	65	7	2.5	51	53	61	68
095	Caramany	42.44	2.35E	300	1	40	190	13	2.6	64	54	63	71
096	Saint Priest	45.58	3.43E	591	2	220	400	13	3.0	103	95	103	110
097	Le Four	45.47	2.36E	725	1	43	95	9	3.2	120	118	128	137
098	Saint Leger La M	46.01	1.25E	576	1	24	170	4	3.4	104	107	117	126
099	Ussel	45.28	2.22E	608	2	45	115	4	3.3	103	98	109	118
100	Cerilly	46.39	2.53E	264	1	90	340	2	3.2	102	102	111	119
101	Massiac	45.15	3.13E	644	1	56	145	13	3.2	130	107	118	127
102	Puy Mary	45.05	2.41E	1095	1	50	210	14	1.9	134	105	112	119
103	Saint Cere	44.50	1.55E	420	3	60	95	1	2.8	50	50	61	69
104	Sainte Clair de	48.41	.36W	230	0	280	385	4	3.0	72	72	80	87
105	Ronchamp	47.43	6.36E	350	4	100	310	7	2.6	108	108	116	123
106	Eschau	48.30	7.45E	144	2	800	850	4	1.8	98	98	101	103
107	Ville	48.21	7.19E	400	8	100	300	11	2.5	97	98	106	113
108	Saint Yriex	45.34	1.11E	388	1	28	111	4	3.3	109	109	119	128
109	Etrez	46.21	5.10E	230	1	1400	1500	4	2.5	102	102	102	101
110	Saint Avold	49.09	6.39E	360	1	150	220	1	4.0	62	62	72	82
111	Cattenom	49.26	6.14E	180	1	50	190	2	1.6	77	77	84	90
112	Besancon Thize	47.15	6.05E	320	1	90	135	1	2.4	74	74	83	91
113	Saint Die	48.19	7.04E	600	1	150	250	2	2.5	88	90	97	104
114	Feldkirch	47.51	7.16E	230	1	600	1600	0	2.5	86	86	88	90
115	Leymen	47.30	7.28E	410	1	500	1120	3	2.5	70	70	74	78
116	La Godardiere	46.59	1.00W	100	2	30	160	7	3.2	63	63	73	81
117	Champrobert	46.55	3.56E	517	1	53	124	4	5.6	73	73	86	98
118	Piriac	47.22	2.32W	7	1	90	120	1	3.5	60	60	73	81
119	Cadeyer	44.02	3.56E	335	1	90	120	3	5.0	102	102	117	128
120	Largentiere	44.34	4.19E	240	1	70	220	7	2.5	80	80	88	96
121	Semalens	43.35	2.04E	240	1	80	250	9	1.6	45	45	51	57
122	Villeveyrac	43.27	3.35E	35	1	100	200	2	2.7	93	93	102	109
123	Randels	44.15	3.03E	865	1	200	230	3	1.8	70	83	89	95
124	Saint Saturnin	44.25	3.03E	655	1	80	420	4	2.5	104	103	111	118
125	Lodeve	43.42	3.22E	139	1	200	250	3	2.0	96	96	102	109
126	Requista	44.01	2.37E	380	3	100	170	4	3.5	108	95	106	115
127	Neffies	43.32	3.21E	190	1	80	350	14	2.5	57	58	66	71
128	Guipy	47.13	3.35E	240	1	50	240	20	3.0	103	103	112	121
129	Muret	43.27	1.22E	175	1	140	180	4	2.4	84	84	91	99
130	Montagne de Blon	46.01	1.00E	410	1	50	400	8	3.2	103	97	106	114
131	La Touche	45.46	.12E	127	1	160	330	5	2.1	70	70	76	83
132	Champ de Moussy	45.46	.09E	130	1	350	450	5	2.5	93	93	100	106
133	Montereau	48.23	2.55E	50	1	150	575	23	1.9	69	69	75	80

Tab. 1 (continued)

NO	Station	LAT	LONG	ALT	N	ZMIN	ZMAX	NM	KM	FX1	FX2	FX3	FX4
134	Manosque	43.52	5.46E	477	1	410	440	9	4.4	100	100	109	118
135	Pechelbronn	48.52	7.55E	150	4	200	600	0	2.5	121	121	128	134
136	Saunay	47.35	.55E	128	1	720	780	3	2.4	85	85	90	94
137	Beauvin	48.37	.19W	260	1	60	143	5	3.3	47	47	59	67
138	Fougeres	48.25	1.21W	180	1	30	95	3	3.1	66	66	75	84
139	Esswein	46.42	1.22W	87	1	40	80	3	6.0	85	85	94	106
140	Mesanger	47.25	1.13W	59	1	40	180	5	3.3	71	71	81	90
141	Saint Jacut	47.41	2.12W	51	1	65	101	3	3.9	95	95	108	117
142	La Porte Aux Moi	48.17	2.55W	280	1	103	175	4	5.0	53	53	67	78
143	Bodennec	48.28	3.36W	410	1	113	242	3	4.3	83	83	94	104
144	Pouande	47.56	1.33W	64	1	72	141	4	3.0	36	36	47	55
145	Estables	44.41	3.29E	1200	1	50	170	10	2.5	104	100	109	117
146	Belmont Buss	47.46	5.32E	127	1	50	170	1	3.4	150	150	161	170
147	Chemery	47.23	1.29E	129	1	840	1100	3	2.5	105	105	108	111

Tab. 2. List of data of group C. Same topics as for Tab. 1.

NO	Station	LAT	LONG	ALT	N	ZMIN	ZMAX	NM	KM	FX1	FX2	FX3	FX4
149	Gournay	49.32	2.42E	100	1	700	740	4	2.6	68	68	73	78
150	Saint Illiers	48.59	1.33E	127	1	518	568	4	2.8	63	63	70	76
151	Beynes	48.51	1.51E	106	1	737	766	4	3.0	54	54	60	65
152	Nancy Velaine	48.35	6.22E	450	1	100	520	6	2.5	72	72	79	86

Origin and classification of the data

168 data have been compiled using published documents (Hentinger & Jolivet 1970, Gable 1977, Groupe Fluxchaf 1978, Gable & Goguel 1979, Gable 1979, Gable & Watremez 1979, Gable 1980, Vasseur 1980), and also unpublished material. Because of the large variety of conditions under which data were acquired, it was decided to separate the data into four groups, according to the characteristics of the borehole (depth range used, thermal equilibrium, temperature measurement) and also according to the way of obtaining the heat flow value (conductivity measurement or estimated values). These four groups, labelled A, B, C and D, are described below:

Group A (90 data) – These data, published by Gable (1977, 1979), are heat flow estimates using file data from the oil industry. The geothermal gradient is appraised from the difference between the ground temperature and a temperature obtained inside the borehole, either during formation testing or during bottom testing (in the latter case the temperature has to be corrected). The thermal conductivity is estimated from lithological logs by using a table of physical constants.

Group B (56 data) – These data are derived from temperature measurements at thermal equilibrium in relatively shallow boreholes (100 to 300 m, generally mining exploration boreholes), and from thermal conductivity measurements on rock samples (Hentinger & Jolivet 1970, Groupe Fluxchaf 1978, Gable & Watremez 1979, Gable 1980, Vasseur 1980). In Tab. 1 an updated list of these classical measurements specifies the geographical coordinates, the depth interval of the measurements, the number of conductivity measurements and several calculations of heat flow (uncorrected and corrected).

Group C (4 data) - These data were obtained in relatively deep boreholes
(500 to 1000 m) by the same method as for data of group B (Gable 1980,
Vasseur 1980). The major difference is that thermal equilibrium is not
ascertained because the boreholes are close to geological synclines used
by the national gas company (Gaz de France) for the storage of gas. This
storage results in a local temperature perturbation which slowly propagates in the surroundings of the reservoir; if heat is transferred only
by conduction, this perturbation can be neglected at a distance of a few
tens of metres. Since some doubts about the occurrence of other transport
phenomena remain, these four data are classified separately and given in
Tab. 2.

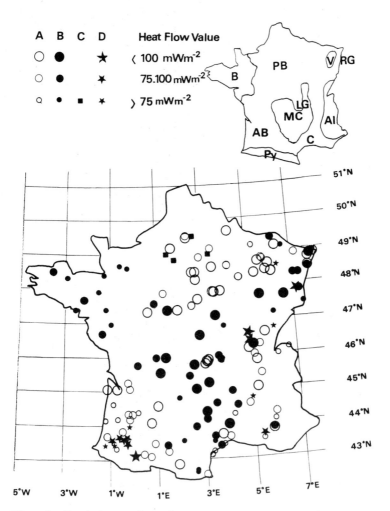

Fig. 1. Positions of various heat flow data. Different symbols are used
for different groups of data (A, B, C, D) and the size of each symbol
depends on the heat flow value (no palaeoclimatic correction was applied).
In the upper map, the main geographical units quoted in the text are
specified (B: Brittany, PB: Paris Basin, AB: Aquitaine Basin, MC: Massif
Central, V: Vosges, C: Camargue, RG: Rhine graben, LG: Limagne graben,
Al: Alps, Py: Pyrenees).

Group D (16 data) - As for group A, these data originate from deep oil exploration boreholes (Gable 1980). In this case, however, several temperatures were obtained at several depths, allowing a local temperature gradient calculation. Moreover, cores were obtained at depth for conductivity determination.

The position of the data is illustrated in Fig. 1. It is evident that data of group A and D cover the deep sedimentary basins (Paris and Aquitaine basins) whereas most data of group B are spread over basement areas (Massif Central, Brittany and Vosges); the 4 data of group C are located in the Paris Basin. The simultaneous use of all these data permits a good coverage of the territory.

Corrections

The apparent heat flow values described in the previous paragraph have to be corrected for several effects such as irregular topography effect, palaeoclimatic influence, erosion and sedimentation effects. For orogenic areas (Alps, Pyrénées) recent rates of erosion and sedimentation are low. Therefore the only corrections considered here are the topographic and palaeoclimatic corrections.

Topographic corrections may be important in the case of shallow boreholes in regions with uneven ground surface. Most of the data of group B have been corrected by using the procedure of Bullard (1940) (also described by Kappelmeyer & Haenel 1974). In a few cases (data no. 137 to 144) the actual position was not known with sufficient accuracy to perform this correction. For data of groups A, C and D, for which the temperature is obtained at greater depth and in relatively flat areas, the topographic correction was neglected.

Palaeoclimatic corrections require the knowledge of palaeotemperature variations. In oceanic sediments as well as in ice shields, temperature variations during the last glacial and interglacial periods can be traced out by using oxygen isotope measurements. However, over continents it is

Tab. 3. Palaeotemperature model for palaeoclimatic corrections.

Episode	Begin. (aB.P.)	End (aB.P.)	Temp. Differences (°C)		Reference
Wurm 1	65000	40000	-6.0	(-12)	Woillard, 1979
Wurm 2	40000	30000	-4.0	(-10)	
Wurm 3	30000	10500	-8.0	(-14)	
Boreal	9000	8000	.5		Lamb, 1977
Atlantic	8000	5000	1.8		
Subboreal	5000	3250	.5		
Subatlanticum	3250	1600	-.4		
Little Clim Opt	1000	700	.8		
Little Ice Age 1	540	470	-.4		Legrand, 1979
Optimum 1530	470	410	.1		
Little Ice Age 2	410	320	-.3		
Little Ice Age 3	320	270	-.2		
Minimum 1800	190	140	-.2		
Optimum 1940	50	20	.4		

necessary to use indirect evidence given by palynology, palaeoecology and
other disciplines (see Lamb 1977). In France, a major source of climato-
stratigraphic information was the detailed pollen study of La Grande Pile
(Vosges) (Woilard 1979); the last postglacial warming can be placed at
11 000 a B.P. and the last glacial period began around 65 000 a B.P. with
a climax period starting around 30 000 a B.P. The amplitude of the postglacial warming is in fact not known; it is generally estimated to be
around 10 °C, but smaller and larger amplitudes have also been proposed.
We therefore have chosen two extreme models: a "low" hypothesis corresponding to a postglacial temperature rise of 8 °C, and a "high" one
corresponding to 14 °C.

The postglacial temperatures may be characterized by several episodes
which are very well described by Lamb (1977): Boreal, Atlantic, Subboreal
and Little optimum. Using Lamb's description, based mainly on biological
evidence for Western Europe, we approximate the corresponding fluctuations
by a succession of steps described in Tab. 3. For the historical period,
namely the Little Ice Age, detailed information from the study of the wine
harvest series (Leroy Ladurie 1967, Legrand 1979) leads to a model with 4
episodes. Finally the 20th century is characterized by a relatively warmer
period culminating around 1940.

The two resulting temperature models corresponding to the two assumptions
for postglacial warming are approximated by a series of step functions as
shown in Fig. 2a. In a homogeneous isotropic semi-infinite medium, the
perturbation of the steady state due to such a series of step functions
may be obtained by using a classical solution to the one-dimensional heat
transfer equation where only the thermal diffusivity occurs (Carslaw &
Jaeger 1959). An example of the resulting corrections on temperature,
temperature gradient and heat flow for a medium with a thermal conductivity
of 2 W m^{-1} °C^{-1} and a diffusivity of 10^{-6}m^{2}s^{-1} is given in Fig. 2b and 2c.
The correction on temperature is positive and at a depth of 1200 m reaches
2.5 to 5 °C according to the palaeoclimatic model. The heat flow correction
is positive near the surface with a value of 8 to 15 mW m^{-2} and becomes
slightly negative below 1200 m.

The two possible heat flow corrections have been applied to all the
data of groups A, B, C, and D, using each of the two hypotheses for the
postglacial rise. A constant heat capacity per volume of 2.10^6 J m^{-3} °C^{-1}
was assumed and, for each borehole, a mean diffusivity was obtained from
the average thermal conductivity measured in that borehole (Beck 1977).
For data of groups B and C, the corrected values are given in Tab. 1 and 2.

Consistency of the data

Individual data are affected by noise resulting from measurement errors
and also from other processes not taken into consideration by the previous
corrections. Data no. 131 and 132 (Tab. 1) offer a striking example: two
values of 73 and 90 mW m^{-2} have been obtained at an interval of only a few
km. For each set of data, a rough estimate of this error may be gained
from the mean square deviation of nearby data: for a distance less than
20 km this deviation reaches 21 mW m^{-2} for data of group A and 19 mW m^{-2}
for data of group B. Such a high value can be attributed partly to in-
accuracy of measurement (for data of group B, the scatter of individual
heat flow measurements generally exceeds 10 %) and also to local perturbations of heat flow due to various phenomena such as water circulation
and conductivity inhomogeneities.

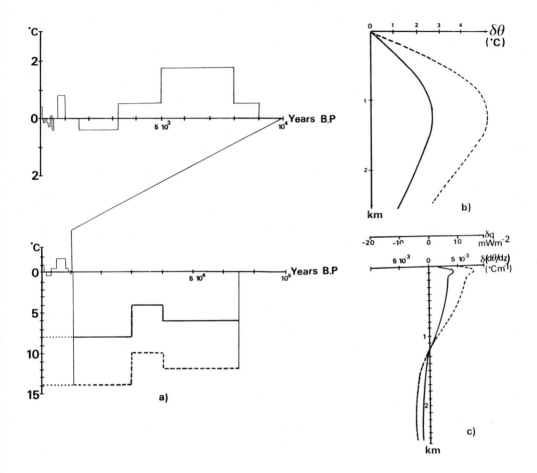

Fig. 2. a) Palaeotemperature model. The temperature deviation (with respect to the present temperature) is plotted as a function of time. The two curves correspond to two assumptions for postglacial warming. The upper part is an enlarged view of the more recent temperature variation (Holocene).
b) Temperature correction as a function of depth (to be added to the observed temperature profile). Conductivity: $2 \text{ W m}^{-1} \text{ }^{\circ}\text{C}^{-1}$; diffusivity: $10^{-6} \text{m}^2\text{s}^{-1}$. The two curves correspond to the two palaeotemperature models.
c) Correction for temperature gradient and heat flow as a function of depth (same model as above).

Another critical discussion is concerned with the mutual consistency among the various groups of data. On some occasions it is possible to compare data which belong to different groups and whose distances do not exceed 30 km. This comparison leads to the result that data of group A generally give values higher by 10 - 50 mW m^{-2} than those of groups B or C. This feature is very clear for Paris basin (see Fig. 1), where a systematic positive difference of 30 - 50 mW m^{-2} is obtained between 9 measurements of group A and the corresponding nearby measurements of groups B or C. If the palaeoclimatic correction is taken into account this difference is only slightly reduced. This large discrepancy has to be

attributed to some other cause which has not yet been clarified. Since
data of group A result from heat flow estimates without actual temperature
gradient and conductivity measurements, a systematic bias on one or both
quantities seems to be a likely explanation.

Tentative heat flow map

As discussed in the previous paragraph, the estimated error for the data
(20 mW m^{-2}) is comparable to the expected amplitude of the regional heat
flow variation. Therefore only smooth estimates can lead to significant
variations; the simultaneous use of several individual data is required.
Moreover, the evaluation of heat flow in a given area depends critically
upon the selection of the data.

    In view of the previous discussion it is reasonable to retain only data
of groups B, C and D, which are considered as more reliable.

    Palaeoclimatic corrections also have some effect on the heat flow
pattern; this is a result of the depth variation of this correction: for
shallow measurements which lie mainly on basement outcrops, this correction
is a positive one, whereas it is negative for deeper determinations.
Therefore a palaeoclimatic correction corresponding to the 8 °C postglacial
warming - which is a likely temperature model - was applied.

    The corresponding heat flow values are plotted in Fig. 3. In order to
point out the most significant heat flow variations, contour lines
accounting for the main trends of these variations have been drawn in Fig.
3. The main features of this map are: the presence of a wide positive

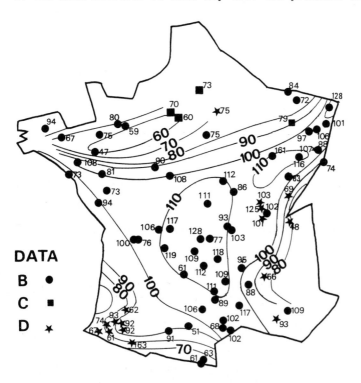

Fig. 3. Tentative heat flow map. The plotted data have been selected
(groups B, C, D) and corrected for palaeoclimatic effects.

anomaly over the Massif Central, with an extension toward the northeast
and the south, and the occurrence of a relatively low heat flow axis from
Brittany to the Paris basin.

In Fig. 4, this tentative map is compared with a simplified geological
map of France and also with maps of the Mohorovičič discontinuity depth
(Hirn 1980) and of the lithospheric thickness deduced from surface wave
dispersion (Souriau 1978).

The main high heat flow anomaly seems to be related to Hercynian
structures and in particular to the presence of granites associated with
the Hercynian orogeny. However, a possible connection of high heat flow

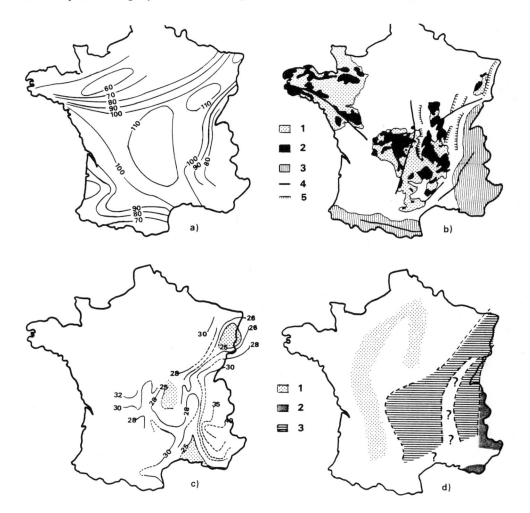

Fig. 4. a) Heat flow map (cf. Fig. 3).
b) Geological scheme showing: 1 = Hercynian basement areas, 2 = granitoids,
3 = Alpine orogenic zones, 4 = main faults, 5: grabens.
c) Moho isobaths in km, deduced from experimental seismology (after Hirn
1980).
d) Lithospheric thickness as deduced from surface wave dispersion (after
Souriau 1978). 1: thickness larger than 100 km, 2: thickness around
100 km, 3: thickness between 60 and 100 km.

with Oligocene grabens and neogene volcanism cannot be excluded; in fact the data are too sparse to allow a more local comparison.

The crust-mantle interface (Fig. 4c) is characterized by a significant uplift in the east (Rhine graben), in the centre (Limagne graben) and in the south (Camargue). A corresponding thinning of the lithosphere (Fig. 4d) is noticeable although it is not so well defined and seems to extend toward the western part of Massif Central. Obviously there is some similarity between the pattern of the positive heat flow anomaly (Fig. 4a) and the area of thin lithosphere. Therefore it is suggested that the high heat flow anomaly is the result of two components: a crustal component originating from heat producing Hercynian granites and a deep component related to the lithospheric thinning.

Conclusion

The various heat flow data for the France are discussed in order to specify the regional trends of heat flow and their possible relations with the main geological structures. These data are supplemented and corrected for topographic and palaeoclimatic effects. A critical study of data consistency is then carried out; consequently, a group of data obtained from heat flow estimates in oil exploration boreholes (group A) is neglected. Using the remaining data, a tentative heat flow map showing interesting relations with the near-surface geology and deep structures is proposed.

A main feature of this map is the occurrence of a positive heat flow anomaly in the Massif Central, which may be related to the presence of Hercynian granites in the crust and/or to some other phenomenon associated with deeper lithospheric structures.

Heat production measurements were recently carried out in the various units of the Massif Central; no clear relation between surface heat production and local heat flow was evident. This result would support the existence of a deep origin for at least a part of the anomaly.

Acknowledgements. This work was supported by the Institut National d'Astronomie et de Géophysique (Action Thématique Programmée "Transfert d'Energie Thermique") and by the Commission of the European Communities (contract no. 482.78 EGF). We are indebted to Dr. Jolivet (Paris), Dr. Meunier (Strasbourg), Dr. Fontaine (Clermont Ferrand), Mr. Maechtelinck (Paris), and Mr. Gable (Orléans) for providing their new data. We are also grateful for the continuous aid of Mr. Kast for the data reduction and of Mrs. Dubuc for translation and typewriting.

References

Beck, A.E., 1977: Climatically perturbed temperature gradients and their effect on regional and continental heat flow means. - Tectonophysics 41: 17-40.

Bullard, E.C., 1940: The disturbance on the temperature gradient in the Earth crust by inequalities of height. - Mont. Not. Astr. Soc. Geophys., suppl. 4: 300-362.

Carslaw, H.S. & Jaeger, J.G., 1959: Conduction of Heat in Solids. - 2nd Ed., Oxford Univ. Press, London, 510 pp.

Gable, R., 1977: Température et flux de chaleur en France. - Seminar on Geothermal Energy, Bruxelles, CEE Editions, 111-131.

Gable, R., 1979: Draft of a geothermal flux map of France. - In: V.
    Čermák & L. Rybach (Eds.), Terrestrial Heat Flow in Europe. - Springer,
    Berlin: 179-185.
—, 1980: Heat flow pattern in France. - 2nd Seminar on Geothermal
    Energy, Strasbourg (to be published, CEE Editions).
Gable, R. & Goguel, J., 1978: Carte du flux géothermique de la France. -
    Compt. Rend. Acad. Sci. Paris 288 D: 195-198.
Gable, R. & Watremez, P., 1979: Premières estimations du flux de chaleur
    dans le Massif Armoricain. - Bull. BRGM Sect. 17: 35-38.
Groupe Fluxchaf, 1978: Nouvelles déterminations du flux géothermique en
    France. - Compt. Rend. Acad. Sci. Paris 286 D: 933-936.
Hentinger, R. & Jolivet, J., 1970: Nouvelles déterminations du flux géo-
    thermique en France. - Tectonophysics 10: 127-146.
Hirn, A., 1980: Le cadre structural profond d'après les profils seismi-
    ques. - In: A. Autran & J. Dercourt (Eds.), Evolutions géologiques en
    France. - Mém. B.R.G.M., Fr, no. 107.
Kappelmeyer, O. & Haenel, R., 1974: Geothermics with special reference to
    application. - Geoexploration Monographs Ser. 1, No. 4, X, 238 pp.,
    Gebr. Borntraeger, Berlin, Stuttgart.
Lamb, H.H., 1977: Climate: Present, Past and Future. Vol. 2: Climatic
    History and the Future. - Methuen and Co. Ltd., London, 333 pp.
Legrand, J.P., 1979: Les variations climatiques en Europe occidentale
    depuis le Moyen Age. - La Météorologie, VIᵒ série, 16: 167-182.
Leroy Ladurie, E., 1967: Histoire du climat depuis l'an mil. - Flammarion,
    Paris, 366 pp.
Souriau, A., 1978: Le manteau supérieur sous la France et les régions
    limitrophes au Nord. - Thesis, University P. et M. Curie, Paris.
Vasseur, G.: A critical study of heat flow data in France. - 2nd Seminar
    on Geothermal Energy, Strasbourg (to be published, CEE Editions).
Vasseur, G. & Nouri, Y., 1980: Some trends of heat flow in France. -
    Tectonophysics 65: 209-223.
Woillard, G., 1979: The last interglacial-glacial cycle at Grande Pile,
    in Northeastern France. - Bull. Soc. Belge de Géol. 88: 51-69.

On the low heat flow in the Transylvanian Basin

S. Veliciu and M. Visarion

with 6 figures

Veliciu, S. & Visarion, M., 1982: On the low heat flow in the
Transylvanian Basin. - Geothermics and geothermal energy, eds. V.
Čermák & R. Haenel, E. Schweizerbart'sche Verlagsbuchhandlung,
Stuttgart: 91-100.

Abstract: In order to explain the low heat flow observed in the
Transylvanian Basin, the geological and geophysical data are reviewed. By means of heat flow and heat generation data resonable
geothermal models are given for Transylvania. This intermountain
basin exhibits some features which make it distinct as compared
with other typical ensialic intra-arc basins.

Authors' address: Institute of Geology and Geophysics, 78344 Bucharest, Romania

Introduction

From the geological point of view the Transylvanian Basin, located in the
inner part of the Romanian Carpathian chain, is defined as a structural
post-tectonic element corresponding to a homogeneous and young (Neogene)
area which was subjected to molasse sedimentation (Dumitrescu & Săndulescu
1968). The molasse deposits overlie a folded basement and its post-tectonic
cover.
   Reviewing the main geological and geophysical characteristics of the
post-tectonic intermountain basins (intermountain trough, intra- and
inter-arc basins), Stegena et al. (1975) find the following common
features:
   - Thinner lithosphere; HCL and LVZ at higher position; lower density in
comparison with the average; positive travel-time residuals of teleseismic events.
   - Thinner crust.
   - Sinorogenic and particularly post-orogenic sediments affected less or
not at all by tectonic movements.
   - Sialic basins exhibiting andesitic volcanic activity of compressional
type during the subducting process in the related areas and "interarc-
spreading" basaltic volcanism of extensional type after the subduction
ceased.
   - Low seismic activity for already developed basins.
   - High heat flow.
From the measurements reported by Demetrescu (1973) and Veliciu et al.
(1977) and from the Heat Flow Map of Europe (Čermák & Hurtig 1979), a low
terrestrial heat flow appears typical of Transylvania (Fig. 1), which is
quite puzzling when compared with the remarkable geothermal activity
observed in the Pannonian Basin. The object of the present contribution is
to try to give a more detailed interpretation of this fact.

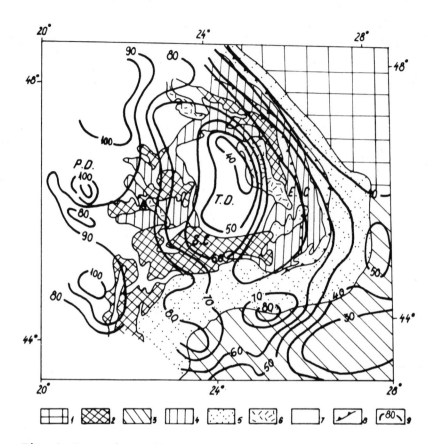

Fig. 1. Tectonic outline and the heat flow of the Transylvanian and related areas after the Heat Flow Map of Europe by Čermák & Hurtig (1979).
1. Precambrian platforms, 2. Hercynian folded units, 3. Epi-Hercynian
platforms, 4. Cenozoic folded units, 5. Alpine foredeeps, 6. Volcanites,
7. Intramountain basins, 8. Thrust faults, 9. Isolines of heat flow
($mW\ m^{-2}$), P.D. Pannonian Basin, T.D. Transylvanian Basin, A. Apuseni
Mountains, E.C. Eastern Carpathians, S.C. Southern Carpathians.

Characteristics of the Transylvanian Basin

Pre-basinal development

The pre-Neogene geological history of the Transylvanian area is closely
related to the evolution of the Carpathians.

In view of petrological evidence Rădulescu & Săndulescu (1973) pointed
to the presence of two intracontinental basins with oceanic-type floor in
the Carpathian area, which were active from the Triassic to the Cretaceous
(Fig. 2). These basins were generated by crustal spreading, probably in
connection with the genesis of the southern border of the Eurasian continent. The expansion ceased by the Aptian-Albian.

The first important compressional deformation due to the approach of
the sialic blocks occurred during the Middle Cretaceous. The western basin
was closed up and the basic rocks together with the accumulated deposits
were bilaterally ejected forming the Transylvanian nappes on the east side

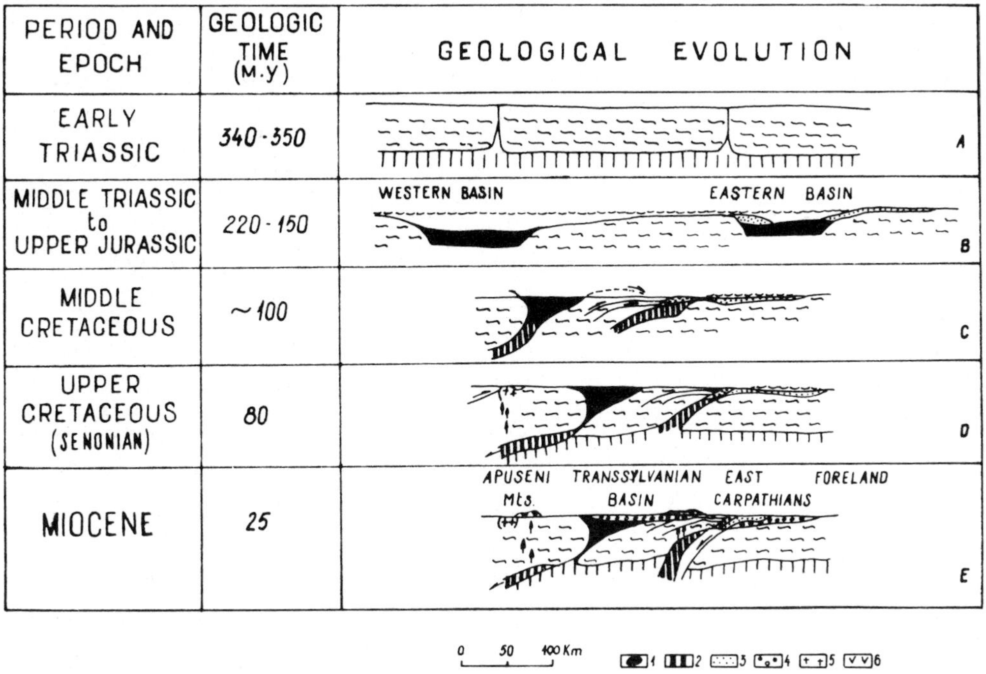

PERIOD AND EPOCH	GEOLOGIC TIME (M·Y)	GEOLOGICAL EVOLUTION
EARLY TRIASSIC	340·350	A
MIDDLE TRIASSIC to UPPER JURASSIC	220·150	WESTERN BASIN EASTERN BASIN B
MIDDLE CRETACEOUS	~100	C
UPPER CRETACEOUS (SENONIAN)	80	D
MIOCENE	25	APUSENI TRANSSYLVANIAN EAST FORELAND Mts. BASIN CARPATHIANS E

0   50   100 Km

Fig. 2. Geological evolution of the Carpathian area after Rǎdulescu & Sǎndulescu (1973). 1. oceanic-type crust, 2. subducted oceanic-type crust, 3. East Carpathians flysh, 4. molasse, 5. Banatitic rocks, 6. Neogene volcanics.

and the overthrust within the Metaliferi Mountains on the west side.

The subduction continued during the late Turonian and Senonian when the nappes were built up in the Northern Apuseni, the Central East Carpathians and in the South Carpathians and Sinaia Beds Zone, respectively. The Banatitic rocks were subsequently emplaced as a result of the subduction of the oceanic-type crust.

The above mentioned processes concluded the first main consolidation stage of the Carpathians. At this point the inner Carpathian zone including Transylvania became stable, whereas the outer zones still remained mobile.

Volcanism

The Carpathian region is characterized by intensive calc-alkaline volcanic activity during the Mio-Pliocene. Around the Transylvanian Basin the areal distribution of volcanism follows the inner boundary of the Carpathian arc and it is made up mainly of andesitic rocks.

Several authors (Rǎdulescu & Sǎndulescu 1973, Pecerillo & Taylor 1976) have suggested that the calc-alkaline volcanism of the area was generated by the subduction of oceanic-type crust; the fragments of the former eastern basin floor were carried in front of the descending continental block.

Structure of the folded basement

From the pre-basinal evolution of the Transylvanian Basin it can be concluded that the main tectogenesis responsible for the structure of its
folded basement took place between the Middle Cretaceous and the Senonian.
    A complex correlation of geological, geophysical and drilling data
reveals that the area occupied by the Neogene molasse of the Transylvanian

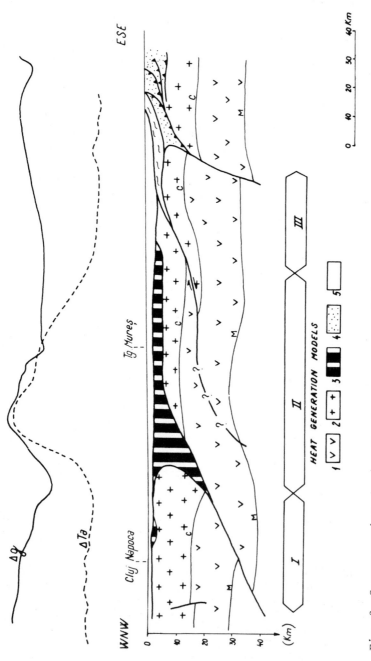

Fig. 3. Cross-section showing the deep structure of the Transylvanian Basin revealed by the
interpretation of DSS and other geophysical data (Rǎdulescu et al. 1976). 1. Basaltic layer,
2. Granitic layer (sialic crust), 3. Transylvanian nappes and South Apuseni (simatic crust),
4. East Carpathian nappe system, 5. Post-tectonic formations. (For heat generation models
see Fig. 5).

Basin covers the junction of structural elements belonging to the major
tectonic units of the Romanian Carpathians (Săndulescu & Visarion 1978).
In the centre of the depression both gravity and magnetic highs are partly
related to the ophiolitic zone which is bilaterally overthrust on the
Apuseni crystalline-bearing basement units westward and on the central East
Carpathian nappe system (the "root" of the nappes) eastward (Fig. 3). The
folded basement is overlapped by a faulted and folded cover of the Neo-
cretaceous, Palaeogene and Lower Miocene age.

The Middle Mio-Pliocene molasse, which is approximately 4 km thick,
fills this intramountain depression.

Crust and upper mantle

DSS data show that the crustal thickness starts with 34 km in the eastern
part of Transylvania and gradually decreases to 30 km in the centre.
Towards the west the crust thickens out again to nearly 38 km. The crust
is relatively thin with respect to the Eastern Carpathians and the foredeep (40 - 50 km), mainly because of the reduction of the upper crust by
6 - 10 km. However, it is thicker in comparison with the crustal thickness in the Pannonian Basin (24 - 28 km).

The vertical distribution of the compressional wave velocity shows the
characteristics of a continental (sialic) type crust even under the
ophiolitic zone. Velocities of $5.5 - 6.2$ km s^{-1} for the upper crust,
$6.8 - 7.0$ km s^{-1} for the lower crust and $8.2 - 9.1$ km s^{-1} beneath the
Moho-discontinuity were recorded (Rădulescu et al. 1976). The last figure
corresponds to a density of the upper mantle which is greater than
average.

As for the lithospheric thickness, no direct investigation has been
carried out so far. Nevertheless, some inferences can be made from the
position of HCL in the eastern part of the Pannonian Basin (Adám 1965)
and in the Vrancea region (Stănică, personal communication) where depths
of 60 km and more than 150 km were recorded, respectively. A depth of
about 100 km for HCL seems to be reasonable under the Transylvanian Basin.

Heat flow and geothermal models of the lithosphere

Surface heat flow

Few heat flow measurements have been conducted in Transylvania (Demetrescu
1973, Veliciu et al. 1977). These data were recently incorporated into
the Heat Flow Map of Europe (Fig. 1) by Čermák & Hurtig (1979). In spite
of the scarcity of data, the regional features of the geothermal regime
of this area become quite clear.

Demetrescu (1973) reported five values measured in the northern and
east central part of Transylvania. Low heat flow values of $33 - 58$ mW m^{-2}
were found. The value of 33 mW m^{-2} seems to be low; the author explains
that the equilibrium time for the borehole before measurements may
possibly have been too short, thus affecting the value obtained. Veliciu
et al. (1977) reported a value of 74 mW m^{-2}, measured near Tîrgul Mures.

However, the fact that the average heat flow in the Transylvanian Basin
is only of $45 - 50$ mW m^{-2} indicates that the geothermal activity of this
area is low. The low heat flow is outlined by the measurements performed
in the surrounding tectonic units, where values of $83 - 110$ mW m^{-2} for the
Neogene volcanic arc of the Eastern Carpathians, of 80 mW m^{-2} for the

South Apuseni and of 85 - 100 mW m^{-2} for the eastern part of the Pannonian Basin were determined (Veliciu & Demetrescu 1979).

Mantle heat flow

The observed continental heat flow can be divided into two components (Pollack & Chapman 1977): one is due to radioactive heat sources in the upper crust, the other to the heat flow contribution from the lower crust and the upper mantle.

The geotherms for a given continental region depend on the surface heat flow, the vertical distribution of radioactive heat sources and the variation of thermal conductivity with depth. Existing experimental data provide indications of the thermal conductivity, but information about the vertical distribution of heat sources is very scanty. Lachenbruch (1970) suggested a decrease of the heat production (A) with depth according to the exponential law.

The linear heat flow vs. heat production relationship, first reported by Birch et al. (1968), offers a reasonable explanation in terms of variations in shallow crustal radioactivity for the range of heat flow within a heat flow province. The relationship is expressed by:

$$Q_O = Q* + bA_O ,$$

where Q_O is the surface heat flow, A_O is the heat production of the surface rocks, $Q*$ is the "reduced" heat flow (the heat flow intercept for zero heat production) and b is a quantity with a depth dimension which characterizes the vertical source distribution.

Pollack & Chapman (1977) established a new empirical equation relating reduced heat flow to mean surface heat flow in a heat flow province ($\overline{Q}_O$):

$$Q* = 0,6 \; \overline{Q}_O .$$

This empirical partition of the surface heat flow thus fixes the contribution of the surface enriched zone at:

$$bA_O = 0.4 \; \overline{Q}_O .$$

If b is known, it yields the mean heat production of the enriched zone. For Transylvania it was assumed that b = 9.8 km. Inserting an average surface heat flow of 50 mW m^{-2} into the calculation, we obtained values of 2.04 mW m^{-3} and 30 mW m^{-2} for A_O and $Q*$, respectively.

The reduced heat flow ($Q*$) is made up of a radiogenic contribution from the lower crust and the upper mantle plus the deeper contribution which arises from the asthenosphere and enters the lithosphere at its base. It is clear that the lower crust is less endowed with heat-producing radioactive sources than the upper crust; on the other hand, it is not known whether a decrement of surface enrichment is accompanied by a proportional decrement or increment in the lower crust. Petrological arguments are consistent with either situation. The problem is open to discussion especially for the ophiolitic zone in the central part of Transylvania, where the heat generated in the upper crust is lowered by a factor of 6 or so, because of the content of basic rocks in the folded basement (Model II in Fig. 5).

For the geothermal models adopted a uniform heat production value of 0.25 µW m^{-3}, which is representative of a granulite-facies in the lower crust, was chosen. Beneath the crustal-mantle boundary a depleted ultrabasic zone with a heat production of 0.01 µW m^{-3} was considered. Both chosen values fit the graph reported by Rybach (1973) according to the heat production-seismic velocity relationship in the range 5.0 to 8.5 km s⁻¹ (Fig. 4).

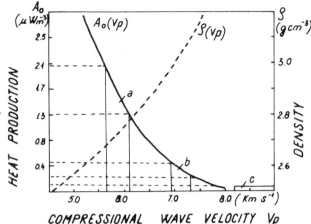

Fig. 4. Heat production vs. seismic wave velocity relationship after Rybach (1976). Compressional wave velocities observed in the Transylvanian Basin: a – upper crust; b – lower crust; c – beneath the Moho.

Heat generation models were established separately for the Southern Apuseni crystalline-bearing basement units, the ophiolitic zone and the central East Carpathian nappe system (Fig. 5). For each model the mantle heat flow (Q_M) can be calculated according to:

$$Q_M = Q. - \int_{o}^{z_{Moho}} A(z)\,dz.$$

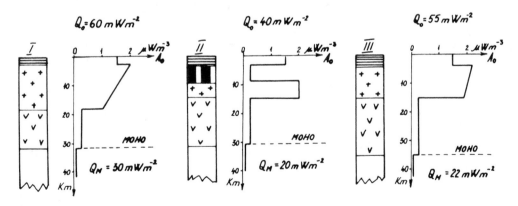

Fig. 5. Heat generation models A(z) for the crustal structure of the Transylvanian Basin. (Model I for the Southern Apuseni crystalline-bearing basement units, Model II for the ophiolitic zone, Model III for the central East Carpathian nappes system).

The calculated mantle heat flow varies from 20 mW m⁻² to 30 mW m⁻² and is close to the reduced heat flow obtained by means of the linear relationship between surface heat flow and heat production.

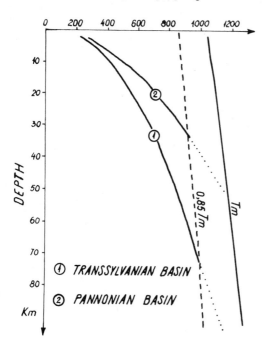

Fig. 6. Temperature vs. depth for Transylvania and the Pannonian Basin. T_m is the mantle solidus. Data for the Pannonian Basin compiled from Horváth & Stegena (1977).

Fig. 6 shows geotherms corresponding to the average surface heat flow observed in the Transylvanian Basin and the Pannonian area (Veliciu 1977). The temperature calculated at the base of the crust in Transylvania (~ 600 $^\circ$C) is lower, as compared to that for the Pannonian Basin (800 – 900 $^\circ$C).

The geotherms were extended to a depth at which they intersect the mantle solidus (T_m) but were dotted to indicate provisionality above 0.85 T_m. The geotherms are characterized by near-surface curvature due to the crustal heat production and a nearly linear gradient through the depleted zone.

Various authors have suggested that the lithosphere-asthenosphere transition might begin at a temperature less than the solidus. Pollack & Chapman (1977) adopted as the depth of the base of the lithosphere the depth at which the geotherms reach 0.85 T_m. The geotherm for the Transylvanian Basin intersects 0.85 T_m at a depth of approximately 80 km while in the Pannonian Basin it reaches the same temperature at a depth of only 40 km. Consequently, the geothermal data clearly indicate a thicker and cooler lithosphere for Transylvania.

Discussion

During the Middle Oligocene-Early Miocene the dipping subduction zone was still active along the Carpathian arc. Presumably the subduction induced the extra heat which resulted in partial melting in the upper mantle above the subducting plate. The hot upwelling mantle material expanded laterally at the base of the continental crust, thinned it out and locally pierced it where the large mass of andesites extruded. When the subduction had ceased, the thin but still continental crust began to subside and was subsequently filled by the Pliocene-Quaternary sediments. This model (Horváth & Stegena 1977) which explains the Late Cenozoic geological history and the geophysical features of the Pannonian Basin quite well, does not entirely fit the characteristics of the Transylvanian Basin.

Owing to the Alpine subduction process in the Carpathians, the Transylvanian area was continuously subjected to compression. There is no evidence indicating an extension similar to that which occurred in the Pannonian Basin after the subduction had ceased. The result is a thicker and cooler lithosphere, a relatively thin crust but with a normally developed lower layer, and the lack of tensional faults and of recent basic magmatic activity within the basin. Consequently, a reduction of the mantle heat flow by a converse phenomenon, as described by Lachenbruch & Sass (1978) for the extensional basins, is very likely to be present here.

A descending convection current in the mantle under the Transylvanian Basin could also be taken into account (Visarion & Săndulescu 1979); this is probably related to the emplacement of the mantle diapir from the Pannonian Basin.

References

Adám, A., 1965: Einige Hypothesen über den Aufbau des oberen Erdmantels in Ungarn. - Geol. Beitr. Geophys. 74 (1): 20-40.

Birch, F., Roy, R.F. & Deker, E.R., 1968: Heat flow and thermal history in New York and New England. - In: Zen, E. et al. (Eds.), Studies of Appalachian Geology: Northern and Maritime. - Interscience, New York: 437-451.

Čermák, V. & Hurtig, E. 1979: Heat flow map of Europe. - In: Čermák, V. & Rybach, L. (Eds.), Terrestrial heat flow in Europe. - Springer Verlag, Berlin, Heidelberg, New York.

Demetrescu, C., 1973: Valori preliminare ale fluxului termic în Transilvania. - St. cerc. geol. geofiz. geogr. Geofiz. 11: 13-21.

Dumitrescu, I. & Săndulescu, M., 1968: Problèmes structuraux des Carpates Roumaines et de leur avant-pays. - Anu. Com. Géol. XXXVI, Bucarest: 195-281.

Horváth, F. & Stegena, L., 1977: The Pannonian Basin: a Mediterranean interarc basin. - In: Biju-Duval, B. & Montadert, L. (Eds.), International Symposium on structural history of the Mediterranean Basins, Split (Yugoslavia) 25-29 October 1976. - Ed. Technip, Paris: 333-340.

Lachenbruch, A.H., 1970: Crustal temperature and heat production: implications of the linear heat flow relation. - J. Geophys. Res. 73: 6023-6029.

Lachenbruch, A.H. & Sass, J.H., 1978: Models of an extending Lithosphere and heat flow in the Basin and Range province. - Geol. Soc. Am., Memoir, 152.

Peccerillo, A. & Taylor, S.R., 1976: Rare earth elements in the East Carpathian volcanic rocks. - Earth Planet. Sci. Letter 32: 124.

Pollack, H.N. & Chapman, D.S., 1977: On the regional variation of heat flow, geotherms and lithospheric thickness. - Tectonophysics 38: 269-296.

Rădulescu, D.P. & Săndulescu, M., 1973: The plate-tectonics concept and the geological structure of the Carpathians. - Tectonophysics 16: 155-161.

Rădulescu, D.P., Cornea, I., Săndulescu, M., Constantinescu, P., Rădulescu, F. & Pompilian, A., 1976: Structure de la croûte terrestre en Roumanie. Essai d'interprétation des études seismiques profondes. - Anu. Inst. Géol. Géophys., Bucharest, L: 5-36.

Rybach, L., 1976: Radioactive heat production in rocks and its relation to other petrophysical parameters. - Pageoph. 114: 309-316.

Săndulescu, M. & Visarion, M., 1978: Considération sur la structure tectonique du soubasement de la Dépression de Transylvanie. - D.S. Inst. Géol. Géophys., Bucarest, LXIV: 153-173.

Stegena, L., Géczy, B. & Horváth, F., 1975: A Pannon-medence késökainozóos fejlödése. - Bull. Hungarian Geol. Soc. 105 (2): 101-123.

Van Eysigna, F.W.B., 1975: Geological time table. - Elsevier Publ. Co., Amsterdam.

Veliciu, S., 1977: Some results on the mantle heat flow investigation in Romania. - Acta. Geol. Scient. Hungaricae 21: 256-258.

Veliciu, S., Cristian, M., Paraschiv, D. & Visarion, M., 1977: Preliminary data of heat flow distribution in Romania. - Geothermics 6: 20-25.

Veliciu, S. & Demetrescu, C., 1979: Heat flow in Romania and some relations to geological and geophysical features. - In: Čermák, V. & Rybach, L. (Eds.), Terrestrial heat flow in Europe. - Springer Verlag, Berlin, Heidelberg, New York: 253-260.

Visarion, M. & Săndulescu, M., 1979: Structura subasmentului Depresiunii Panonice în Romănia (sectorele central si sudic). - St. cerc. geol. geofiz. geogr. Geofiz. 17 (2): 191-201.

Geothermal investigations on the territory of Byelorussia

Gerasim V. Bogomolov, Yury G. Bogomolov, Vladimir I. Zui and
Leo A. Tsybulya

with 2 figures and 1 table

Bogomolov, G.V., Bogomolov, Y.G., Zui, Vl. I. & Tsybulya, L.A.,
1982: Geothermal investigations on the territory of Byelorussia. -
Geothermics and geothermal energy, eds. V. Čermák & R. Haenel,
E. Schweizerbart'sche Verlagsbuchhandlung, Stuttgart: 101-105.

Abstract: Geothermal investigations have been carried out in all
the main geological structures. The results of deep borehole
temperature measurements and the data from temperature logs are
generalized. Heat flow values were calculated from measurements
performed in 35 deep boreholes taking into account the measured
values of temperature gradients and heat conductivities of the
rocks. The relation of the geothermal field conditions to the
geological pattern of some structures is studied. From a geothermal
standpoint the most detailed data are obtained from the oil-bearing
Pripyat depression. The highest heat flow observed here is 109 mW
m^{-2}.

Authors' address: Institute of Geochemistry and Geophysics,
B.S.S.R. Academy of Sciences, Minsk, USSR.

Introduction

Byelorussia is located in the western part of the East-European platform.
The positive structure - the Byelorussian massif (Fig. 1) is surrounded
by: the Orsha (in the east), Pripyat (in the south), and Brest (in the
west) depressions, separated by the Latvian, Polesskaya, and Zhlobin
saddles. The geology of the territory in the country has not been studied
uniformly. The greatest number of deep holes were drilled in the Pripyat
depression. The temperature measurements in the boreholes were conducted
and the geothermal maps of the territory constructed by the Institute of
Geochemistry and Geophysics, and the heat flow was calculated for a
number of stations (Atroshchenko 1975, Bogomolov et al. 1972). The thermophysical properties of rocks were measured by the transient probe method
(Atroshchenko & Zui 1974, Lubimova et al. 1964).

Subsurface temperature data

The direct subsurface temperature data were obtained in many boreholes.
Their positions are shown in Fig. 1. The isotherms on the surface of the
basement are constructed from these results and drawn in Fig. 2. The depth
of the basement varies from 200-800 m for the massif to 5 km for some
lower parts of the Pripyat depression. It does not exceed 2 km for the
Brest and Orsha depressions. The maximum temperatures at the surface of
the basement in the Pripyat, Brest, and Orsha depressions correspond to
125, 40, and 36 °C respectively, and the minima of 10 - 15 °C belong to

Fig. 1. Geothermal study stations in Byelorussia.
I Byelorussian massif, II Latvian saddle, III Orsha depression, IV Pripyat
depression, V Brest depression, VI Polesskaya saddle, VII Zhlobin saddle.
Black points show the stations of geothermal investigations.

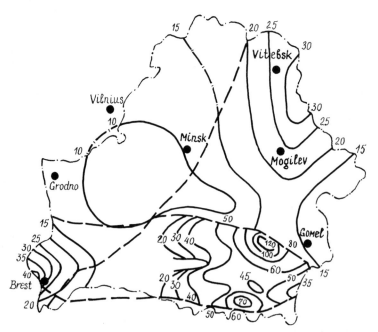

Fig. 2. A schematic map of temperatures at the basement surface. Compiled
by Bogomolov, G.V. & Tsybulya, L.A. Dashed lines show regional fractures.

Tab. 1. Heat flow data from Byelorussia.

No.	Hole		Lati-tude (N)	Longi-tude (E)	Ele-va-tion (m)	Interval of depth (m)	Gra-di-ent (mK m⁻¹)	Heat flow (mW m⁻²)	Refe-rence
Pripyat depression									
1	Berezina	1	52°15'	29°45'	126	941-2475	21	84	1
2	Borisovka	1	52°45'	28°45'	137	646-2250	15	63	1
3	Borshchovka	1	52°20'	30°35'	126	1469-2339	13	71	1
4	Vasilevka	1	52°10'	30°30'	138	1130-2198	13	80	1
5	Visha	15	52°35'	29°15'	145	749-2831	14	71	1
6	Visha	17	52°35'	29°00'	146	690-2877	15	75	1
7	East-Pervomaisk	1	52°30'	30°15'	140	1132-3266	20	84	1
8	Vyshemir	2	52°05'	30°25'	130	1009-1173	13	71	1
9	Glussk	1	52°50'	28°45'	141	540-1720	14	71	1
10	Davydovka	5	52°30'	29°35'	143	868-2699	21	80	1
11	Davydovka	10	52°30'	29°25'	140	889-2646	15	71	1
12	Elsk	21	51°50'	29°05'	152	1100-2353	12	54	2
13	Zolotukha	2	52°15'	29°45'	139	790-1600	13	84	1
14	Malodusha	1	52°10'	30°05'	136	784-1551	15	88	1
15	Malynskaya	3	52°45'	28°50'	145	740-2934	14	71	1
16	Oktyaborsk	3	52°30'	29°05'	130	808-2648	12	67	1
17	Ostashkovichi	1	52°25'	29°55'	N.A.	1020-2881	16	71	1
18	Ostashkovichi	19	52°25'	29°50'	140	1376-2500	17	67	1
19	Pervomaisk	1	52°35'	30°00'	134	997-3466	18	109	1
20	Pervomaisk	5	52°35'	30°00'	139	642-2977	17	92	1
21	Rechitsa	1	52°15'	30°15'	129	415-749	13	88	1
22	Rechitsa	17	N.A.	N.A.	127	526-2673	15	88	1
23	North-Domanovichi	3	52°30'	29°15'	139	648-2476	10	63	1
24	Tishkovka	1	52°25'	30°05'	133	742-1898	18	75	1
25	Hatetskaya	1	N.A.	N.A.	132	540-1310	18	59	1
26	Chervonaya Sloboda	4	52°30'	28°40'	143	677- 800	11	63	1
27	Chernin	1	52°35'	29°15'	146	768-3521	16	80	1
28	Sharpilovka	1	52°10'	30°45'	118	783-1200	20	92	1
29	Shatilki	1	52°35'	29°45'	131	932-3326	20	92	1
30	Shatilki	2	52°35'	29°45'	135	700-1960	21	84	1
Brest depression									
31	No 7		N.A.	N.A.	N.A.	500-1000	20	52	3
32	Brest-OP 1 (NO8)		52°15'	23°30'	156	841-1212	N.A.	59	3
33	No 15		N.A.	N.A.	N.A.	150- 457	20	42	3
Orsha depression									
34	Orsha-OP 2		54°25'	30°40'	197	300- 550	16	34	2
Byelorussian massif									
35	Skidel	1	53°35'	24°10'	110	260- 762	16	39	2

Notes: N.A. = Not available; for references: 1 = Atroshchenko, 2 = New data, 3 = Bogomolov et al.

the Byelorussian massif. The temperatures on its slopes and saddles do not exceed 20 °C. The least values of the geothermal gradients, 8 - 12 mK m^{-1}, observed in sedimentary deposits, is also related to the massif. The temperature inside the massif basement was measured in 12 boreholes. In the Skidel borehole 1 at a depth of 762 m it equals 19.4 °C. Similar values of the gradients in the sediments of the Brest and Orsha depressions were obtained.

The thickness of the sediments of the oil-bearing Pripyat depression varies from 2 to 5 km. Salt tectonics and high conductivity of rock salt (6-8 Wm^{-1}K^{-1}) noticeably influence the nonuniformity of the temperature field. The total thickness of salt reaches 3 km and the basement has a block structure. The temperature at its surface increases in the northward direction, reaching 125 °C in the region of a heat anomaly near the northeastern fault zone, where there is an abrupt basement submersion to a depth of 5 km. The 20 °C isotherm in the Pervomaisk borehole 5 corresponds to a depth less than 200 m, and in the southern and western parts of the depression it reaches 1200 m (in the following boreholes: Elsk-11, Turov-2, Chervonaya Sloboda-1). The geothermal gradients in sedimentary strata are within the limits of 10 - 20 mK m^{-1}.

Heat flow data

The heat flow calculations for the boreholes are performed with the use of the average temperature gradients and mean values of the thermal conductivity. Heat flow data from Byelorussia are summarized in Tab. 1 for 35 deep boreholes. For individual boreholes the previous values of heat flow were re-evaluated in view of the additional data. Some unreliable results are excluded. Most of the heat flow observations were made in the Pripyat depression. The heat flow distribution is nonuniform here. The least value of 54 mW m^{-2} corresponds to the Elsk borehole 21. Anomalous heat flow is displayed near the fault in the northeastern zone. The highest value of 109 mW m^{-2} was obtained in the Pervomaisk borehole 1.

Conclusions

- The complexity of geological and hydrogeological conditions in Byelorussia results in the nonuniformity of its geothermal field.
- An increased heat flow from the basement in the northeastern zone of the Pripyat depression causes more intensive heating of the sedimentary cover here and is connected with a deep fracture.
- The results of the investigations of geothermal activity show the possibility of utilizing thermal water in the Brest and Pripyat depressions for heating greenhouses, buildings, etc., and for the recovery of dissolved chemicals from brines.

References

Atroshchenko, P.P., 1975: Geothermal conditions of the northern part of Pripyat depression. - Nauka i tekhnika, Minsk, 104 pp. (Russ.)
Atroshchenko, P.P. & Zui, V.I., 1974: Scheme of thermophysical properties measurements of rocks, based on the transient heat probe method. - In: Problems of geochemical and geophysical investigations of the Earth's crust. - RISO AN BSSR, Minsk, pp. 185-194. (Russ.)

Bogomolov, G.V., Tsybulya, L.A. & Atroshchenko, P.P., 1972: Geothermal
    zones of the BSSR territory. - Nauka i tekhnika, Minsk, 212 pp. (Russ.)
Lubimova, E.A., Starikova, G.N. & Shushpanov, A.P., 1964: Thermophysical
    investigations of rocks. - In: Geothermal investigations. - Nauka,
    Moscow, pp. 115-174. (Russ.)

Thermal regime and terrestrial heat flow in permafrost areas of the USSR

V. T. Balobaev and V. N. Deviatkin

with 1 figure and 2 tables

Balobaev, V. T. & Deviatkin, V. N., 1982: Thermal regime and terrestrial heat flow in permafrost areas of the USSR. - Geothermics and geothermal energy, eds. V. Čermák & R. Haenel, E. Schweizerbart'sche Verlagsbuchhandlung, Stuttgart: 107-110.

Abstract: The results of the determination of the terrestrial heat flow in Siberia and in the northeast of the USSR by the Geothermal Laboratory of the Permafrost Institute, Siberian Branch, USSR Academy of Sciences are analyzed.

Authors' address: Permafrost Institute, Siberian Branch, USSR Academy of Sciences, Yakutsk, USSR.

Investigations in the permafrost regions have been conducted only during recent years. At present only basic concepts of the geothermics under permafrost conditions in Siberia exist, and general features of the largest geological-tectonic formations are studied. The main formations are as follows: the West-Siberian platform, the Siberian platform with the Anabar crystalline massif, the Aldan shield with the southern delineation of the Baikal orogene region and the Verkhoyano-Kolymskaya mountain folded area.

The West-Siberian platform is tectonically relatively uniform, and therefore both temperature and heat flow hardly vary over the entire area. Permafrost is spread north of $60°$ northern latitude. There is continuous permafrost beyond the Polar Circle, and it is 400-500 m thick.

To the south, the permafrost table is no longer present at the surface and gradually sinks to a depth of 200 m. This frozen layer is a relic of the past cold epoch. The permafrost base gradually rises toward the south to the depth of 200-250 m. It was found that from west to east the temperature falls and the permafrost thickness increases.

Tab. 1. Temperature of the sedimentary cover on the West-Siberian platform, $°C$.

	Temperature at a depth of		
	500 m	1000 m	3000 m
Southern (Ob basin region) (Priobye)	10-20	30-45	95-120 $°C$
Central	8-15	25-35	90-115
Northern (circumpolar)	9-10	29-30	80-90
Yamal peninsular	-	30-35	80-105
Eastern (Yenisei basin region)	6-8	20-25	65-75
North-eastern	1-3	12-15	50-70

The southern part of the area under discussion (Ob Basin region) is the warmest; at a depth of 1000 m the temperatures are 30-45 °C, and at a depth of 3000 m they reach 95-120 °C. Such high temperatures are charact-eristic of the depressions with a thick layer of young, poorly cemented sandy-clayey deposits. The temperature decreases northwards by 5-15 °C at a depth of 1000 m, and by 15-30 °C at a depth of 3000 m (Tab. 1). All over the platform, the heat flow changes within 50-67 mW/m^2 decreasing from the west to the east.

Tab. 2. Temperature and the heat flow of separate structures on the Siberian platform.

Structure	Temperature, °C			Heat flow,
	500 m	1000 m	3000 m	mW/m^2
Anabar crystalline rock mass	-5	-1	12-14	-
Southern slope of Anabar anteclise	-4 - -2	-3 - +3	15-26	17-21
Western slope of Anabar anteclise	-1	7	35-40	21-25
Nepsko-Botuobinskiy arch	-2 - +3	1-5	13-23	13-21
Lena foredeep	2	5	22-24	42
Aldan shield	5-8	12-17	44-50	38-42
Yenisei basin region	5-7	7-15	25-45	21-34
Vilyui anteclise	0-1	12-20	65-73	46-55
Near Verkhoyanskiy Ridge foredeep	-0.5 - +1	11-15	60-66	46-55
Yenisei-Khatang foredeep	-1 - +7	10-20	50-80	34-55
Kitchan cusp	10 - 14	27-30	79-83	60

The vast region of the Siberian platform has not yet been closely studied. The data available show that the underground temperature and heat flow values are the lowest here, and the permafrost thickness is the most highly developed (1500 m). The oldest and elevated part of the platform, composed of Proterozoic and Lower Palaeozoic crystalline rock and carbonate sediments, is notable for the temperatures 12-26 °C at a depth of 3000 m and for very low heat flow (less than 21 mW/m^2 (Tab. 2). This can be explained as follows:
- by prolonged tectonic passivity of this part of the platform,
- by the presence of thick saline surface deposits with a high thermal conductivity and
- by prolonged (hundreds of millions years) surface exposure and intensive cooling in the circumpolar region.
  In the platform depressions with a young sedimentary cover the de-pression is the younger, the higher the temperature and the higher the heat flow. In the youngest of them - Vilyui syncline - the temperature

reaches 60-75 °C at a depth of 3000 m, and the value of the heat flow exceeds 45 mW/m^2 (Deviatkin 1975).

The Aldan and Anabar anticlines are similar in composition, but differ greatly in their thermal state. The near-surface temperature in the Aldan anticline is 2-3 times higher and the heat flow is 2 times higher than in the Anabar anticline. The distribution of permafrost is discontinuous there, and its thickness rarely exceeds 100 m. Following the origin of the Stanovoi Ridge, the southern termination of the Aldan shield was subjected to activation in the Mezo-Cenozoic, and, in our opinion, this led to the increase in thermal activity. In the ridge, the value of the heat flow is approximately 60 mW/m^2.

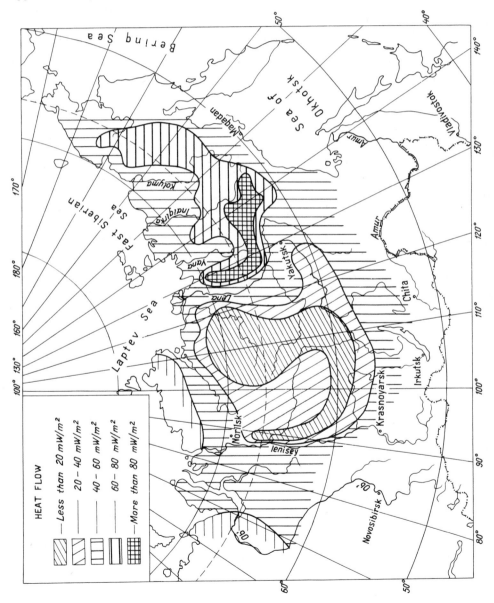

Fig. 1. Map of the terrestrial heat flow in permafrost area of the USSR.

The Verkhoyano-Kolymskaya mountain folded area is the most active geo-tectonic province in the permafrost area. The uplifting of the main mountain systems, which started in the Mesozoic, conditioned the supply of the great amount of heat and for the upward movement of molten rocks. As a consequence, in the Cherskiy and Verkhoyanskiy Ridges the rock temper-ature at a depth of 3000 m reaches 80-100 $^\circ$C, and the heat flow reaches its maximum. In the Suntar-Khayata Ridge, the value of heat flow is 100 mW/m^2.

In the highlands and uplands, the heat flow decreases to 65-80 mW/m^2, and in the lowlands of the northeast and intermountain areas, it becomes equal to 40-65 mW/m^2. The temperatures decrease correspondingly. The high value of the heat flow and considerable thermal conductivity of the rocks are responsible for the relatively small thickness of permafrost. Even at present, at very low surface temperatures (-6 to -8 $^\circ$C), it rarely exceeds 200-300 m. The strong influence of relief determines a very different geo-thermal and geocryological situation in the mountain regions.

According to the data obtained by the Permafrost Institute, a sketch map of the terrestrial heat flow distribution in the permafrost area of the USSR beyond the Urals was compiled (Fig. 1). All the main geotectonic elements are clearly distinguished in the map. As striking features is the contrast between the thermal conditions and the heat flow, observed on crossing the Yenisei tectonic structure, separating the West-Siberian plat-form and the Siberian platform from the west to the east, and when passing from the latter to the orogenic structures in the northeast. At the joint of the West-Siberian platform and the Siberian platform, the heat flow value increases sharply at a distance of 150-200 km, and at a depth of 3000 m the temperature decreases by 30-40 $^\circ$C; the horizontal gradient of temperature exceeds 0.2 $^\circ$C/km. The eastern boundary of the Siberian plat-form is also clearly distinguished from the Verkhoyanskiy Ridge by both temperature and heat flow, but this boundary is poorly defined because of the presence of the Near Verkhoyanskiy foredeep (Tab. 2). This is indica-tive of the occurrence of shallow sources for this significant thermal contrast. In the inner part of the West-Siberian platform and of the de-pressions a great amount of heat seems to evolve from the consolidation of loose deposits as they accumulate. Calculations show that the low heat flow observed in the Siberian platform can be generated by a minimum presence of radioactive elements in the earth's crust, without any other sources and without any heat supply from the mantle.

The high values of heat flow in the Cherskiy and Verkhoyanskiy Ridges are directly related to the process of orogenesis by which, in the final analysis of energetic changes, a great amount of heat is evolved, and these high values of heat flow depend on the intensity of this process at the present epoch.

Reference

Deviatkin, V.N., 1975: The results of the determinations of the terrestrial heat flow in Yakutia. - In: Regional and Thematic Geocryological Studies, Novosibirsk, Nauka, p. 148-150. (Russ.)

The subsurface temperature atlas of the European Communities

Ralph Haenel

with 1 figure

Haenel, R., 1982: The subsurface temperature atlas of the European Communities. - Geothermics and geothermal energy, eds. V. Čermák & R. Haenel, E. Schweizerbart'sche Verlagsbuchhandlung, Stuttgart: 111-112.

Author's address: Geological Survey of Lower Saxony, P.O.B. 510153, D-3000 Hannover 51, Fed. Rep. of Germany

During the first "Research and Development Programme on Geothermal Energy of the European Communities" from 1976 to 1979 temperature maps have been constructed for an area of about 1.5 Mil. km^2. The temperature atlas includes maps in increments of 500 m down to 3000 m and a map of the temperature distribution at 5000 m depth. The atlas consists of:

Maps of Europe, scale 1:5 000 000, and
Maps of E.C. Member States, scale about 1:2 000 000.

For areas which are covered by sufficient temperature data and which are of special interest for geothermal energy exploration additional maps have been presented, such as the so-called:

Regional maps, scale about 1:500 000, and
Local maps, scale about 1:50 000.

The map (Fig. 1) shows the temperature distribution at 1000 m depth below surface for the European Community.
  Furthermore, the atlas contains maps on geology, terrestrial heat flow density, temperature at the earth's surface, interpretation of temperature data in terms of water circulation at depth, etc.
The knowledge of the temperature field in the subsurface is of interest for geosciences, such as:
- geothermics, as a necessity for the determination of the geothermal energy resources and reserves as well as for the exploration of geothermal energy itself;
- oil and gas exploration, because the evolution of oil and gas reservoirs depends mainly on temperature;
- hydrology, since the ascending and descending water in the subsurface is connected with heat transport and therefore detectable;
- geodynamics, because the temperature is an important parameter. For that purpose, a large-scale as well as a homogenous representation is required.
  The temperature atlas is published by Th. Schäfer GmbH, Tivolistraße 4, D-3000 Hannover 1.

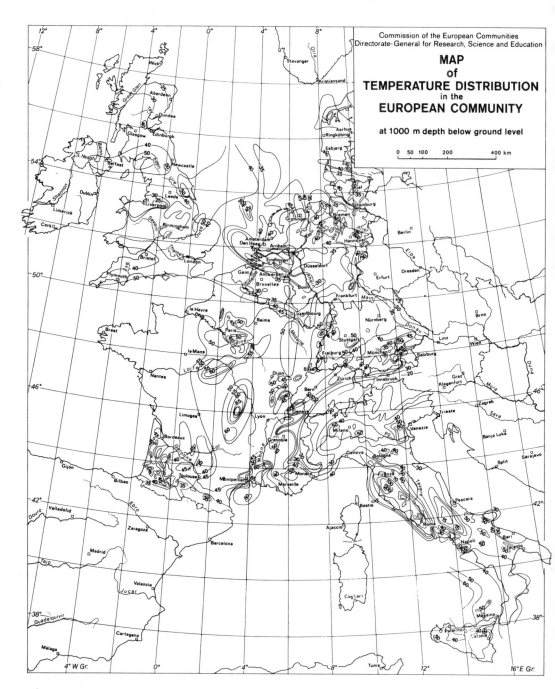

Fig. 1.

The subsurface temperature field of the Federal Republic of Germany

J. Wohlenberg

with 4 figures

Wohlenberg, J., 1982: The subsurface temperature field of the
Federal Republic of Germany. - Geothermics and geothermal energy,
eds. V. Čermák & R. Haenel, E. Schweizerbart'sche Verlagsbuchhandlung, Stuttgart: 113-117.

Author's address: RWTH Aachen, Templergraben 55, D-5100 Aachen,
Fed. Rep. of Germany

This paper represents the extended summary of the already published
collection of maps showing the distribution of the subsurface temperatures
over the territory of the Federal Republic of Germany (Wohlenberg 1979).
The temperature data for these maps were taken from (1) the drilling
records in the archives of the state geological surveys of the Federal
Republic, (2) new temperature logs made by the Lower Saxony Geological
Survey, and (3) heat flow data for greater depths. Temperature data were
available from a total of more than 4800 oil drill holes. In 250 drill
holes high-quality temperature logs were measured and 67 heat flow values
could be used to extrapolate the temperature field to greater depth. These
temperature data were plotted vs. depth for each survey map (1:25 000),
unit areas of 12 km x 12 km. Temperature values were taken from these
graphs for the following depths: 100 m, 250 m, 500 m, 1000 m, 1500 m,
2000 m, 3000 m, 4000 m, and 5000 m. These values were then plotted on a
grid map. A "sliding" average was applied to these values over each group
of nine unit areas for the above-listed depths down to 3000 m. Isotherm
maps were then made from these averaged values. Figs. 1-3 give examples of
isotherm maps of the Federal Republic of Germany for the depths of 250 m,
1500 m and 3000 m. The depth is given in terms of ground level according
to an international recommendation. The maps reveal a rather uniform
temperature distribution with two significant exceptions: The Urach anomaly
SE of Stuttgart and the anomaly of the Rhinegraben structure with its
maximum near Landau half-way from Frankfurt to Freiburg. It has to be
pointed out that the lateral extent of the anomalies is exaggerated as a
result of the smoothing procedure. A net temperature high marks the
sedimentary basin of Northern Germany, striking EW. This temperature
structure was the subject of combined magnetotelluric investigations
(Knödel et al. 1979). Temperature-depth functions for several regions of
the Federal Republic of Germany are given in Fig. 4.

References

Wohlenberg, J., 1979: The Subsurface Temperature Field of the Federal
    Republic of Germany. - Geol. Jb., E 15: 3-29, Hannover.
Knödel, K., Losecke, W. & Wohlenberg, J., 1979: A Comparison of Results of
    Geothermal and Magnetotelluric Investigations in Northwestern Germany. -
    J. Geophys. 45: 199-207.

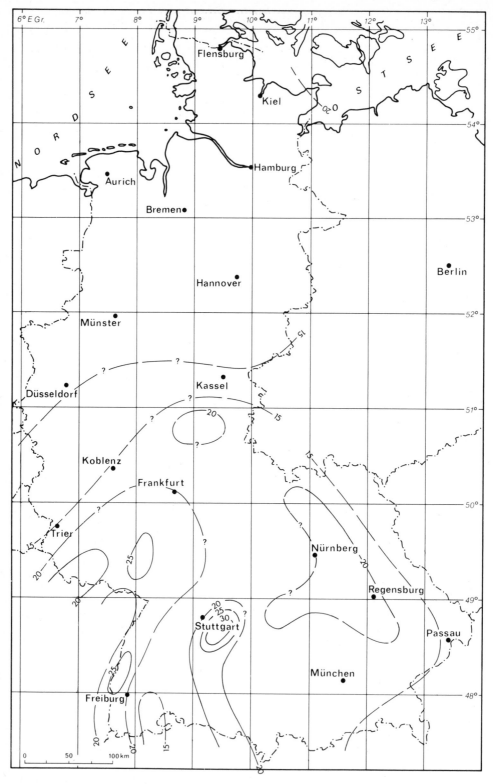

Fig. 1. Isotherm map for a depth of 250 m in the Federal Republic of Germany.

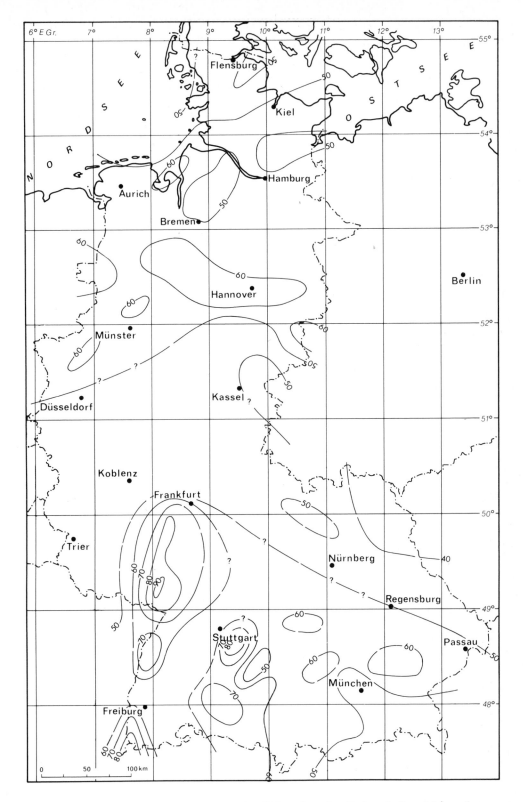

Fig. 2. Isotherm map for a depth of 1500 m in the Federal Republic of Germany.

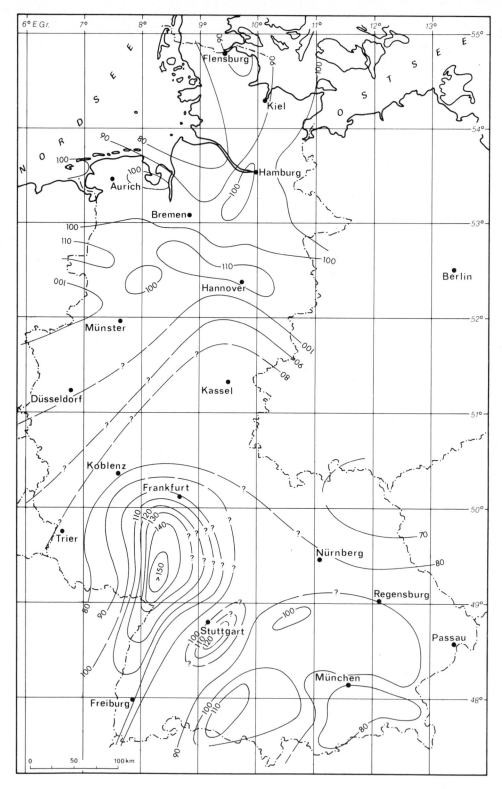

Fig. 3. Isotherm map for a depth of 3000 m in the Federal Republic of Germany.

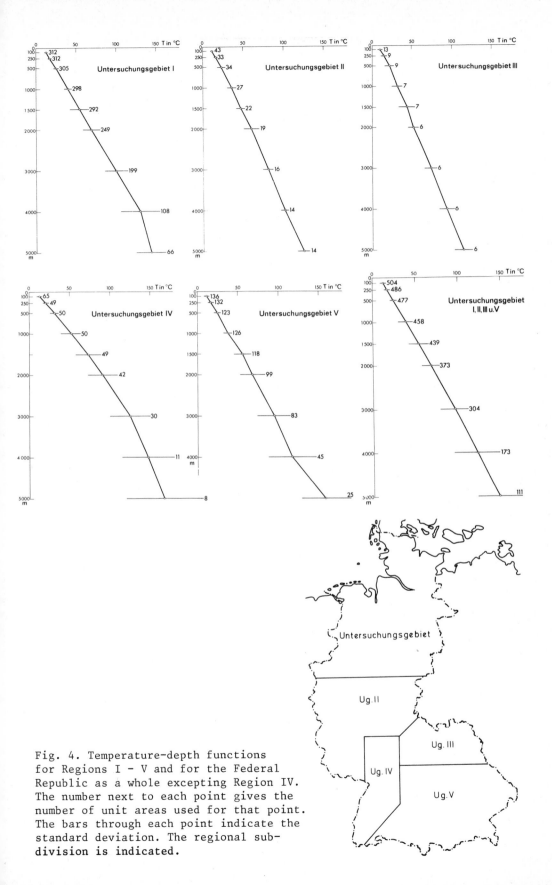

Fig. 4. Temperature-depth functions for Regions I – V and for the Federal Republic as a whole excepting Region IV. The number next to each point gives the number of unit areas used for that point. The bars through each point indicate the standard deviation. The regional sub-division is indicated.

In situ determination of thermal conductivity in boreholes

J. Behrens, B. Roters and H. Villinger

with 3 figures

Behrens, J., Roters, B. & Villinger, H., 1982: In situ determination of thermal conductivity in boreholes. - Geothermics and geothermal energy, eds. V. Čermák & R. Haenel, E. Schweizerbart'sche Verlagsbuchhandlung, Stuttgart: 119-123.

Abstract: A new method for the in situ determination of thermal conductivity in cased or uncased water-filled drillholes is presented. The method uses the principle of an infinite cylindrical heat source. A separated section of the water-filled drillhole is heated at a constant rate of heat input. The temperature rise of the well-mixed water section, measured as a function of the heating time, allows the determination of the thermal conductivity. Assumptions and advantages of the method used are discussed. The construction of the probe is described and first results of in situ measurements of the thermal conductivity are presented.

Authors' address: Institut für Angewandte Geophysik, Technische Universität, D-1000 Berlin 12

Introduction

For heat flow investigations the knowledge of the thermal parameters of rocks, mainly the thermal conductivity, is indispensable. For measuring the thermal conductivity of rocks two different methods exist: The first is to measure the thermal conductivity on rock samples, usually cores, with the aid of steady state or instationary laboratory methods, if cuttings or cores are available. The second is to determine the thermal conductivity in situ in drillholes. The advantages of an in situ determination are obvious. Therefore the development of a suitable in situ probe was necessary. The construction and testing of a probe for the in situ determination of thermal conductivity in cased drillholes are described in this paper.

Thermal conductivity determination

Following the basic investigations of Beck et al. (1956) the measuring principle consists of a cylindrical source heated in a drillhole. The source is realized as a separated section of a water-filled cased or uncased drillhole. This method is basically similar to the well known needle-probe method (von Herzen et al., 1959).

If the length-to-diameter ratio of the cylinder exceeds a value of about 20, and the surrounding rock has a uniform and isotropic thermal conductivity, the heat source may be assumed to be an infinite line source. According to this theory, the temperature rise measured by means of a temperature sensor at the centre of the line source is a linear function of the logarithm of the time of heating. The slope of that

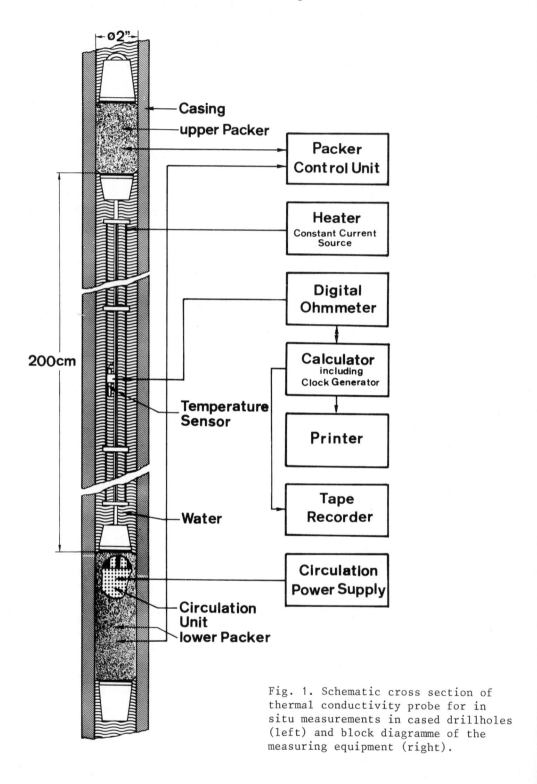

Fig. 1. Schematic cross section of thermal conductivity probe for in situ measurements in cased drillholes (left) and block diagramme of the measuring equipment (right).

straight line is inversely proportional to the thermal conductivity of the rock.

The realization of the measuring principle is shown schematically in Fig. 1. To separate a section of the water-filled drillhole, two inflatable packers are used; these operate with air pressure or on the basis of the bottle brush principle. The thermally insulated water column in this section is heated by a constant current input to four heating wires. To fulfill the condition of a perfectly conducting probe (Beck et al. 1956) the water column is agitated with the aid of a small pump installed in the lower packer. Because of the mixing it is sufficient to measure the temperature in the centre of the column. The temperature sensor is a calibrated thermistor.

The recording and control unit is located at the earth's surface. It consists of a unit for handling and control of the packers, the constant current source for heating, and the device for measuring and storing the data with the aid of a computer controlled data logging system.

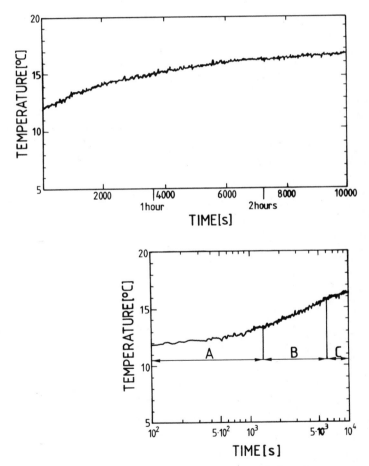

Fig. 2. Example of an in situ measurement of thermal conductivity in a cased drillhole.
The temperature rise of the water column as a linear function of time is shown above, whereas the lower part shows the linear relationship between the temperature rise and the logarithm of time after some time of heating.

Results

Up to now, 17 successful in situ determinations in shallow, PVC-cased
drillholes have been performed. The surrounding rock consists mostly of
limestone or claystone. An example of a measurement is shown in Fig. 2.
The total heating time amounts to about three hours. Three parts of the
curve can be distinguished:
Part A (about 20 minutes) is necessary to heat the water and the probe.

Part B (about 70 minutes) is strongly influenced by the thermal parameters of the casing and the annular space.

Part C (about 90 minutes) is used for the determination of the thermal
conductivity of the rock.

The computation of the slope of the straight line was done with a least
squares procedure. The fit of the data is remarkably good.
    To test the reliability of the in situ values thermal conductivities of
core samples from the same depth interval were measured in the laboratory.
Fig. 3 shows the comparison of in situ values and laboratory values. The
in situ values agree remarkably well with the laboratory values; the
deviations are mostly within a range of ± 10 %. This comparison can give
only a very rough idea of the accuracy of the method because of the
different spatial dimensions, in which the thermal conductivities in situ
and in the laboratory are measured.

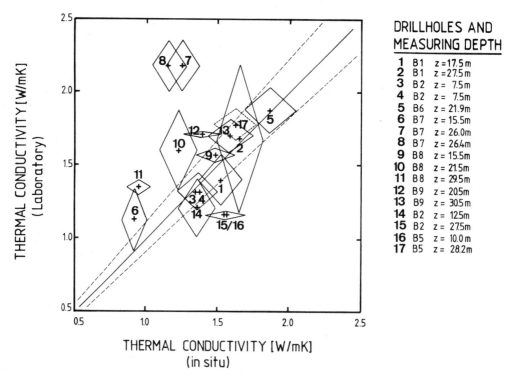

Fig. 3. Comparison of thermal conductivities measured in situ and in the
laboratory.
The straight line marks the points of equality of in situ and laboratory
values and this dashed line shows the error interval of ± 10 %.

Conclusion

The first results obtained with the new probe are very encouraging. The advantages of the proposed method can be summarized as follows:
- The probe has a simple mechanical construction and is therefore very suitable for rough field conditions.
- The probe can be adapted very easily to various diameters of drillholes.
- Measurements are possible in cased or uncased drillholes.
- The thermal conductivity is measured in the same spatial dimensions as the temperature gradient. It therefore represents an average over a reasonably large volume of rock.

Errors due to the annular space, the casing and the finite length of the probe can be eliminated by means of a numerical simulation of the whole system which is in progress.

Acknowledgement: The authors are indebted to Dr. Haenel and Mr. Zoth, both from Niedersächsisches Landesamt für Bodenforschung (Hannover, FRG) and Mr. Geiser, Mr. Schenkluhn and Mr. Schmarsow also to Mr. Töpper and Mrs. Cramer (all TU Berlin) for their help and advice. The work was kindly supported by the Commission of the European Communities (Brussels) and the Ministry of Research and Technology of the Federal Republic of Germany.

References

Beck, A.E., Jaeger, J.C. & Newstead, G., 1956: The measurement of the thermal conductivities of rocks by observations in boreholes. - Australian J. Physics 9: 286-296.

Behrens, J., Roters, B. & Villinger, H., 1980: In situ determination of thermal conductivity in cased drillholes. - Proceedings of Seminar on Geothermal Energy: 525-534, Straßburg.

von Herzen, R. & Maxwell, A.E., 1959: The measurement of thermal conductivity of deep-sea sediments by a needle-probe method. - J. Geophys. Res. 64: 1557-1963.

Measurements of the thermal conductivity of rocks under natural conditions

R. I. Kutas, M. I. Bevzyuk, O. A. Gerashchenko and T. G. Grishchenko

with 2 figures

Kutas, R.I., Bevzyuk, M.I., Gerashchenko, O.A. & Grishchenko, T.G.,
1982: Measurements of the thermal conductivity of rocks under
natural conditions. - Geothermics and geothermal energy, eds. V.
Čermák & R. Haenel, E. Schweizerbart'sche Verlagsbuchhandlung,
Stuttgart, 125-128.

Abstract: The technique of the thermal conductivity determining of
rocks in situ is based on the recording of heat flow disturbances
simulated by an alien body of known thermal conductivity inserted
into the borehole. A high-sensitivity thermoelectric heat flux
sensor is used as such a body. At a given depth in the hole, the
temperature gradient is measured simultaneously in the rocks and
in the sensor. The thermal conductivity coefficient of rocks is
calculated from the ratio of the gradients and the known value of
the thermal conductivity of the sensor.

Authors' addresses: R. I. Kutas and M. I. Bevzyuk, S. I. Subbotin's
Institute of Geophysics, Academy of Sciences of the Ukrainian SSR,
Kiev, U.S.S.R.; O. A. Gerashchenko and T. G. Grishchenko, Institute
of Engineering Thermophysics, Academy of Sciences of the Ukrainian
SSR, Kiev, U.S.S.R.

Introduction

While studying the thermal conductivity of rocks, the investigators have
hitherto limited themselves to small-scale measurements of the core
samples collected from boreholes. The accuracy of these measurements
(Gerashchenko & Grishchenko 1978) may not meet the requirements for
determining the true in situ value of the coefficient of thermal conductivity. The samples extracted from the rock may differ their physical
properties from the same rock under natural conditions.
    When the samples are prepared for measurement, the rock structure and
humidity are disturbed, the samples are very often completely destroyed,
and the measurements are carried out under different thermodynamic conditions. All these limitations have called for the development of a
method of determining the thermophysical properties of the rock directly
in the borehole.
    Investigation in this field has been conducted only for a relatively
short period of time. Several methods of measuring the thermal conductivity
of the rock in situ have been proposed (Filippov 1964, Beck et al. 1971).
They are based mainly on the study of the nonstationary thermal field
caused by drilling in the borehole or created artificially after the
drilling has been terminated.

Method of measuring thermal conductivity in situ

The method presented in the paper for measuring the thermal conductivity of the rock under natural conditions is based on the study of the earth's thermal field and the disturbance of same, as induced by inserting a body with a known thermal conductivity into the borehole. This method is believed to yield more reliable results (Bevzjuk et al. 1980).

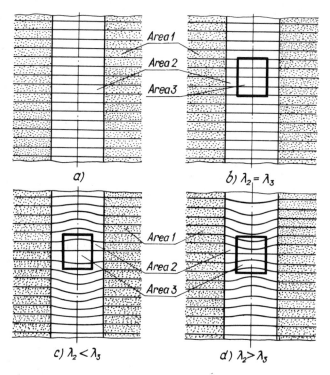

Fig. 1. The character of isotherm distribution in the borehole.

For a borehole with steady-state thermal conditions, it has been stated that the geothermal gradient in the rock massif (area I, Fig. 1) and in the circulating fluid (area 2, Fig. 1) are the same when a body of finite dimensions is introduced into the borehole (area 3, Fig. 1). The temperature field changes according to the ratio of thermal conductivities on the fluid and the probe inserted. If their thermal conductivities are equal, i. e. $\lambda_2 = \lambda_3$ then the temperature field is not disturbed and the geothermal gradient is equal to that of the probe (see Fig. 1b). If the thermal conductivity of the probe is higher or lower than that of the fluid, then the gradient in the probe decreases or increases, respectively, as compared with the original geothermal gradient (see Fig. 1b, d).

Thus the presence of a body of finite dimensions in the borehole practically does not disturb the geothermal gradient in the region near the borehole, but does cause a change in the local temperature gradient in the zone of probe insertion (area 3). The ratio $N = \gamma_r/\gamma_s$ of gradients in the rock massif and in the probe inserted is inversely proportional to the ratio of their thermal conductivities $L = \lambda_r/\lambda_s$.

The functional dependence N = f(L) is constructed on the basis of the results of calculation and simulation with due consideration given to the shape and size of the reference probe placed in the protective shell, this dependence being the calibration curve. Fig. 2 shows such a curve obtained by simulation and verified by the results on thermal conductivity of the rock as determined in the borehole and in the laboratory on the samples taken from the same depths. If this dependence is used, and the thermal conductivity of the reference probe is known, the thermal conductivity coefficient of the rock can be calculated as $\lambda_r = L \cdot \lambda_s$.

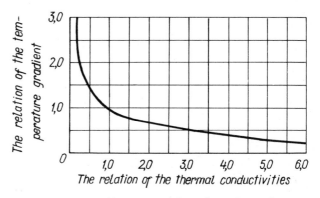

Fig. 2. The function of the temperature gradient relation (geothermal γ_r and in the probe γ_s) depending on the relation of the thermal conductivity (of the rocks λ_r and probe λ_s).

The application of the method is reduced to measuring the geothermal gradient in the rock massif and the temperature gradient in the reference probe. The difficulties of recording very small temperature changes (not higher than 10^{-4}K) in the reference probe can be overcome with the help of a highly sensitive transducer (heat flux meter) with a sensitivity of about 1 mV·m^2/W (Gerashchenko 1971). From the heat flux value measured with the help of the heat flux meter, the temperature gradient is determined, N is calculated and L is found by the calibration curve.

The time required for determining the thermal conductivity of the rock by this method depends on the time needed for the establishment of stationary thermal conditions in the borehole after insertion of the probe; it takes about 1.5-2 hours. The relative error in determining the thermal conductivity of the rock is the sum of the errors incurred in measuring the geothermal gradient, the heat flux density and the thermal conductivity of the heat flux meter, and does not exceed 8 per cent.

References

Beck, A.E., Anglin, P.M. & Sass, I.H., 1971: Analysis of heat flow data – in situ thermal conductivity measurements. – Canad. J. Earth Sci. 8, No. I: 1-19.

Bevzjuk, M.I., Gerashchenko, O.A., Grishchenko, T.G. & Kutas, R.I., 1980: New method of determination of the thermal conductivity of rocks in boreholes. – Promyshlennaya teplotehnika, No. 4: 99-102 (Russ., with Engl. abstract)

Filippov, P.I., 1964: On the technique of determination of thermophysical
    properties of rocks in boreholes. - In: F.E.Arre (Ed.), Thermal Pro-
    cesses in Permafrost Rocks. - Moscow, Nauka, 178-180. (Russ.)
Gerashchenko, O.A., 1971: Fundamentals of Thermometry. - Kiev, Naukova
    Dumka, 192 pp. (Russ.)
Gerashchenko, O.A. & Grishchenko, T.G., 1978: Heat metering method and
    device for thermal conductivity investigation. - In: 6th European
    Conference, Thermophysical Properties of Materials - Research and
    Application. Abstracts, Book I. Dubrovnik, Yugoslavia.

Multifactor dependences of thermophysical properties of rocks

Pavel P. Atroshchenko, Yury G. Bogomolov and Mikhail D. Parkhomov

with 1 figure

Atroshchenko, P.P., Bogomolov, Y.G. & Parkhomov, M.D., 1982:
Multifactor dependences of thermophysical properties of rocks. -
Geothermics and geothermal energy, eds. V. Čermák & R. Haenel,
E. Schweizerbart'sche Verlagsbuchhandlung, Stuttgart: 129-132.

Abstract: With the use of the experimental data on thermophysical
properties of the Pripyat depression sediments, the heat conductivity regression lines are calculated for terrigenous, carbonaceous,
and salt deposits while taking into account their physical and
mineralogical characteristics. The regression equation is obtained
by linearizing the generalized Lichtenecker equation for each type
of rock.

Authors' address: Institute of Geochemistry and Geophysics,
B.S.S.R. Academy of Sciences, Minsk, U.S.S.R.

Introduction

Experimental data on rock thermophysical properties show a complicated
behaviour with respect to their dependence on many characteristics (structural, textural, mineralogical, etc.) and physical parameters, corresponding to bedding conditions (Dortman 1976). Most of the one-dimensional
empirical dependences are of qualitative character. Therefore their usefulness for the calculation of thermophysical properties is very limited.
Multidimensional dependences of heat conductivity and thermal diffusivity
can be used in limited intervals of parameter values for the stochastic
models examined (Anand et al. 1973, Chudnovskii 1976, Tikhomirov 1968).
The multifactor dependences considered in this paper were found for the
Pripyat depression sediments on the basis of experimental data on rock
samples of different petrographical composition. The thermophysical properties of 500 samples with their natural water content were measured.
They were taken from depths of 100-4000 m. The data on the temperature
dependence of thermophysical properties were obtained from 25 samples in
the interval of 20-100 °C. The data on the petrographical, granulometric,
mineralogical composition and physical properties are taken from Makhnach
et al. (1966).

Multiple regression for heat conductivity of terrigenous sediments

The main thermophysical characteristics - volume heat capacity, heat
conductivity, and thermal diffusivity - are interdependent. Two of them
must be determined independently. The volume heat capacity of a sample,
as an extensive quantity, has an additive property. It can be obtained as
a sum of the volume heat capacity of phases. The coefficient of heat
conductivity, which is a function of the given parameters, is used as the
second independent characteristic. This set of parameters is chosen from

the condition of their independence. The common porosity, temperature, content of sand, quartz (for terrigenous rocks), clay fraction (for carbonaceous and salt deposits) are considered as independent parameters.

The analytical expression for the effective heat conductivity K_e is obtained from the Lichtenecker equation (1926); for a two-phase medium it is written

$$K_e = K_1^m \; K_2^{1-m},$$ (1)

where K_1, K_2 = phase heat conductivities; m = volume part of the first phase. For rocks of multicomponent composition, in particular for terrigenous rocks, equation (1) can be modified to yield

$$K_e = ((K_q^q \; K_p^{1-q})^s \; K_g^{1-s})^{1-p} \; K_f^p ,$$ (2)

where K_q, K_p, K_g, K_f = mineralogical heat conductivity coefficients of quartz, feldspar group minerals, clay minerals, and fluid in porous medium, respectively; q = content of quartz in sandy-aleurite fraction; s = content of sand in rocks; p = common porosity. The constants of equation (2), representing the average coefficients of heat conductivity of group minerals and its parameters q, s, and p are stochastic values. The parameters K_e, calculated from (1) or (2) may differ considerably from experimental values.

A more precise definition of the constants in equation (2) can be realized by statistical methods. By expanding (2) the Taylor series around $q = 0$, $s = 0$, $p = 0$ in succession and neglecting quadratic and higher and triple combinations we obtain

$$K_e = K_g + sK_g \ln \frac{K_s}{K_g} + qsK_g \ln \frac{K_q}{K_s} + spK_g \ln \frac{K_s}{K_g}(\ln \frac{K_f}{K_g} - 1) + pK_g \ln \frac{K_f}{K_g} .$$ (3)

Expression (3) represents a linearized form of equation (2) and includes mixed effects of parameter interaction. Let us admit that mixed effects are new independent parameters. The series of dimensionless parameters varying within the limits 0 - 1 in (3) must then have the factor properties. If the addends containing the constants in (3) are denoted as $C^o{}_i$ and the corresponding factors are denoted as t_i, this yields

$$K_e = \sum_i C^o{}_i \; t_i.$$ (4)

Here, $C^o{}_o$ = generalized heat conductivities of clay minerals, and $t_o = 1$. Expression (4) provides a basis for composing an algebraic equation system of the regression analysis. It can be solved in terms of unknown constants C_i. The dispersion matrix analysis, obtained from the matrix of factor combination sums, shows that there exists a statistical relationship between some factors of equation (4). This circumstance renders it necessary to seek the solution of the system of equations in terms of C_i by means of the iteration method of Zeidel by algorithm

$$C_i^{(n+1)} = (\sum_k Kt_i - \sum_{j=0}^{i-1} C_j^{(n+1)} \sum_k t_i t_j - \sum_{j=i+1}^{m} C_j^{(n)} \sum_k t_i t_j) / \sum_k t_i^2$$ (5)

Here, n denotes the nth step; m = quantity of C_i coefficients; $\sum_k$ = summation over all the experimental values of K.

Results

A zeroth-order approximation of the regression equation for the coefficient
of heat conductivity for terrigenous rocks has the form

$$K_e = 2.09 + 0.548 \ s + 2.04 \ qs - 1.33 \ sp - 2.97 \ p \pm 0.56, \tag{6}$$

where K_e is expressed in $Wm^{-1}K^{-1}$. It received on the basis of equation (3)
by means of the values of mineralogical heat conductivities of individual
phases (Subotin & Kutas 1974). The correction of coefficients obtained
from equation (5) gives a better approximation to experimental data.
Finally the regression of the heat conductivity coefficient was obtained
in the form

$$K_e = 1.98 + 0.852 \ s + 2.31 \ qs - 0.98 \ sp - 2.23 \ p \pm 0.33 \tag{7}$$

The remainder dispersion of equation (7) lies between the dispersion of
experimental reproducibility of the heat conductivity coefficient determination and the dispersion of the stratum heat conductivity values. It
offers the possibility of satisfactorily describing the heat conductivities
of terrigenous rocks by (7) with coefficients defined more precisely.
    The temperature dependence of the relative heat conductivity thus
obtained for terrigenous rocks having different porosity (8) can be
neglected for the conditions of the Pripyat depression. See Fig. 1.

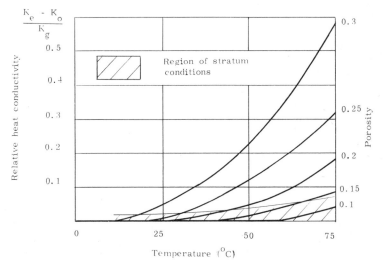

Fig. 1. The dependence of the relative heat conductivity $(K_e-K_o)/K_g$ of
terrigenous rocks on porosity and temperature.

$$\frac{K_e - K_o}{K_g} = 4.04 \ (pT)^{2.8} \cdot 10^{-5} \tag{8}$$

where K_o = heat conductivity at $T = 0 \ ^oC$. The influence of temperature is
evident only at a high degree of rock porosity.
    The procedure described is applied to find the best approximation for
the heat conductivity coefficients of carbonaceous and salt rocks. The

equations of heat conductivity coefficients for dolomite (9) and limestone (10) have an analogous form.
Only the numerical values of the coefficients differ:

$$K_d = 6.07 - 6.45 \ g + 16.75 \ gp - 26.04 \ p \tag{9}$$

$$K_1 = 3.48 - 4.07 \ g + 14.74 \ gp - 8.08 \ p \tag{10}$$

where g = dimensionless clay content of rock.
The temperature is the main factor affecting the heat conductivity of pure salt rocks. Therefore, the equations are the simplest for rock salt (11) and anhydrite (12).

$$K_s = 7.33 - 2.80 \ T \cdot 10^{-2} \tag{11}$$

$$K_a = 5.65 - 1.76 \ T \cdot 10^{-2} \tag{12}$$

Here, K_d, K_p, K_s, K_a are expressed in $Wm^{-1}K^{-1}$.

Conclusions

The relationships presented for heat conductivities of main rock types of the Pripyat depression are in good agreement with the experimental data. Their analytical representation permits the calculation of the thermophysical properties of stratigraphic subdivisions of a complex composition by using the known physical conditions of their bedding and the geological character of deposits. The system of equations of heat conductivity coefficients thus obtained represents the basis for a mathematical model of the geotemperature field in the Pripyat depression.

References

Anand, J., Somerton, W.H. & Gomaa, E., 1973: Predicting thermal conductivities of formations from other known properties. - Soc. Petr. Engrs. J., 13, No 5: 267-273.
Chudnovskii, A.F., 1976: Thermophysics of soils. - Nauka, Moscow, 13-15. (Russ.)
Dortman, N.B. (Ed.), 1976: Physical properties of rocks and fossils (Petrophysics). Geophysical manual. - Nedra, Moscow, 256-280. (Russ.)
Lichtenecker, K., 1926: Die Dielektrizitätskonstante natürlicher und künstlicher Mischkörper. - Physik. Z. 27: 115-158.
Makhnach, A.S., Korzun, V.P., Kurochka, V.P., Laputz, V.A., Uryev, I.I. & Shevchenko, T.A., 1966: Lithology and geochemistry of Devonian sediments of the Pripyat depression with regard to their oil possibilities. - Nauka i tekhnika, Minsk, 314 p. (Russ.)
Subotin, S.I. & Kutas, R.I. (Eds.), 1974: Terrestrial heat flow in the European part of the USSR. - Naukova dumka, Kiev, 20-23. (Russ.)
Tikhomirov, V.M., 1968: Conductivity of rocks and its relationship with density, saturation and temperature. - Neftyanoe Khozyaistvo, 46, No 4: 36-40. (Russ.)

Thermal properties of the lithospheric mineral matter under high pressure and temperature

T. S. Lebedev, V. I. Shapoval and A. A. Pravdivy

with 3 figures

Lebedev, T.S., Shapoval, V.I. & Pravdivy, A.A., 1982: Thermal
properties of the lithospheric mineral matter under high pressure
and temperature. - Geothermics and geothermal energy, eds. V.
Čermák & R. Haenel, E. Schweizerbart'sche Verlagsbuchhandlung,
Stuttgart: 133-139.

Abstract: It is shown that in complex studies of the thermal properties of rocks under the experimental conditions of high pressure and temperature, the methods of the regular thermal regime
are the most appropriate. At the Department of Physical Properties
of Mineral Matter of the Earth, Institute of Geophysics, Ukrainian
Academy of Sciences, Kiev, special devices with various measuring
cells were designed to be used in the study of thermophysical parameters of mineral matter under varying experimental regimes of P-T
conditions.
The results of the temperature conductivity studies on some rocks
in the Ukrainian Shield under high hydrostatic pressure up to
10 kbar are presented. The work reveals a specific behaviour of
this parameter under the effect of varying pressure. Three zones
of temperature conductivity variation are distinguished. For the
first time, an attempt is made to explain the temperature conductivity behaviour within each of the three zones.

Authors' address: S. I. Subbotin's Institute of Geophysics,
Academy of Sciences of the Ukrainian SSR, 32 Palladin av.,
252164 Kiev, U.S.S.R.

Introduction

The insufficient amount and low accuracy of the data on the behaviour of
the thermal conductivity (λ), temperature conductivity (a) and heat
capacity (C) of the minerals under varying pressure (P) and temperature
(T) introduce errors into the solution of the problems of the physics of
the earth and of applied geothermics.

Consequently, great importance is attached to the investigation into
the thermal properties of rocks and minerals under experimental conditions
of high P-T regimes (Lyubimova et al. 1964, Magnitsky et al. 1971, Stiller
et al. 1972, Seipold & Gutzeit 1977, Shapoval & Pravdivy 1978, Shapoval
et al. 1978, Shapoval et al. 1978a, Lebedev et al. 1979).

The modern methods of studying the thermal parameters of mineral matter
under atmospheric pressure and at room temperature have proved inappropri-
ate for high P-T experiments, because of certain restrictions in the
chamber volume and the samples studied, the number of electric leads, the
duration of the experiment, the stability of the properties of the
pressure conducting media, etc. Therefore, the choice of a thermophysical
measurement technique based on solving the heat equation with certain

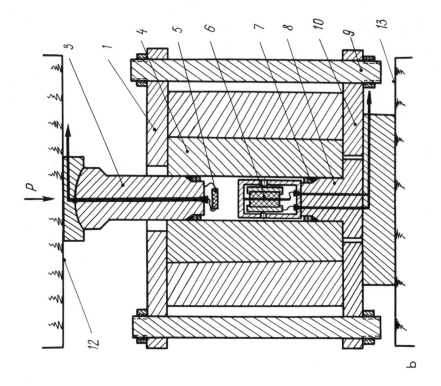

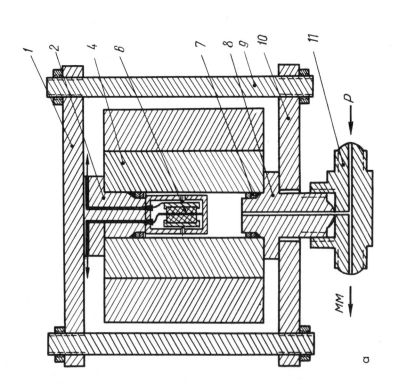

Fig. 1. Designs of high pressure chambers for thermophysical measurements. a – with high hydrostatic pressure generator, b – with multiton hydraulic press. 1 – upper plug of chamber, 2 – upper plug, 3 – upper mobile plug, 4 – two-layered high pressure chamber, 5 – manganine manometer chamber, 6 – measuring cell, 7 – packing, 8 – lower plug, 9 – column, 10 – lower plate of chamber, 11 – T-joint (P – from pressure generator, MM– to manganine manometer), 12 – upper plate of press, 13 – lower plate of press.

boundary conditions is of primary importance (Lyubimova et al. 1964, Yurchak et al. 1973, Lebedev et al. 1979, etc.). It has been shown in many publications that in studying the thermal characteristics of mineral matter under high P and T, the methods based on the theory of the regular regime of the 3rd kind were the most promising (Magnitsky et al. 1971, Yurchak et al. 1973, Shapoval & Pravdivy 1978, Lebedev et al. 1979, etc.). Their major advantage is the possibility of manifold determinations of thermal properties during one experiment. The larger amount of information thus derived (compared to the other methods) and a variety of ways of determining the thermal parameters of minerals make it possible to check the results of the experiment and hence to increase their reliability.

Instruments and technique

The experience gained in studying the elastic, magnetic and electrical characteristics of mineral matter under high P and T conditions (Lebedev 1975, Lebedev et al. 1976, etc.) has made it possible to develop devices and techniques for studying the thermal properties of rocks and minerals under varying P-T regimes, including the methods of the regular regime of the 3rd kind with the application of internal heaters (Shapoval & Pravdivy 1978, Shapoval et al. 1978a, Lebedev et al. 1979).

In a special chamber for high (up to 15 kbar) hydrostatic pressure the thermal properties of the mineral samples can be examined. Because of the design of the vessel it can be used either with or without a high pressure generator (Fig. 1, a); in the latter case high pressure is generated by means of a hydraulic press (Fig. 1, b). The main unit of the vessel is a two-layered self-adjusting cylinder with an inner diameter of 32 mm. The inside of the vessel is made of hardened steel with high tensile strength. The working space of the vessel sealed at the top and bottom with compressing plugs. The construction of the plugs differs depending on the technique adopted for high pressure generation. When a special generator of high hydrostatic pressure is used, the lower plug is provided with a bore to let the pressure transferring liquid pass and is T-joined to the generator and a manganine manometer (Fig. 1, a). All the electric leads for measuring thermal parameters are mounted in the upper plug. When high pressure is generated by a hydraulic press, the measuring cell is fixed on the lower plug, in which the electric leads are mounted. The plug remains immobile during the experiment. To generate high pressure the vessel is placed between the plates of the press (serving as a multiplier) that pushes the upper mobile pumping plug (Fig. 1, b).

The methods of plane and radial thermal waves are employed in these experiments. Accordingly, two different cells have been developed (Lebedev et al. 1979). Both cells are used with the same measuring instruments (Fig. 2). In the feeding unit of the device a thermal wave is periodically generated by the heater element (plate or linear). For this purpose, sub-sonic signals are fed by the stabilized power supply through the master stage to the heater. The forming pulses corresponding to the on- and off-positions of the feeding current are recorded simultaneously with the temperature variation on the sample surface. Thus, one record contains information on both the thermal state of the object studied and the initial conditions of the pulses. The information is subsequently used in calculating thermal characteristics.

The measuring unit consists of a thermocouple, compensator, direct current amplifier and recorder (Fig. 2). In all the experiments thermo-couples with a sensitivity of 25 mV/$^{\circ}$C are used. By means of the commutator

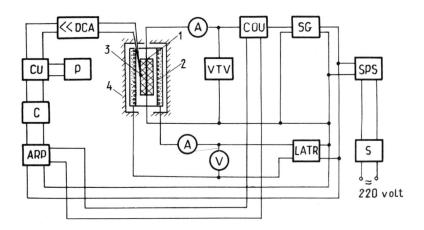

Fig. 2. Device for determining thermal properties of mineral matter under high experimental P-T regimes. 1 - heater, 2 - furnace, 3 - thermocouple, 4 - high pressure chamber, DCA - direct current amplifier, CU - commutation unit, ARP - automatic recording potentiometer, P - potentiometer, C - compensator, SG - subsonic generator, COU - control unit, A - ampere-meter, V - voltmeter, LATR - laboratory transformer, SPS - stabilized power supply, S - stabilizer, VTV - vacuum-tube voltmeter.

the signal is passed either directly to the recorder or through a compensator and amplifier. In the first case, the total temperature of the sample required for reducing the measurement to conditional units is recorded. In the second case, the constant part of the temperature can be balanced with the aid of a compensator and only its variable part, which originate in the sample due to the thermal wave, is amplified and recorded.

Results

The first results of the experiments with the use of the instruments and technique described are shown in Fig. 3. The rocks studied are medium-grained and porphyry-like biotite granite, fine-grained hypersthene charnokite, biotite garnet granite, medium-grained diorite of biotite amphibole composition, coarse-grained labradorite, medium-grained pyroxenite sampled from separate small bodies, eclogite-like rocks of granulite texture, biotite and amphibolic gneiss, all sampled from the Ukrainian Shield.

  The behaviour of the temperature conductivity coefficient (a) under changing pressure has proven to vary from rock to rock. In charnokite, some samples of granite and labradorite (Fig. 3, a, b) the temperature conductivity increases slightly with pressure.

  In diorite, pyroxenite, eclogite-like rocks and in a number of equigranular granite samples (Fig. 3 a, c), the temperature conductivity increases rapidly. All the curves show a temperature conductivity, which changes with pressure in a non-linear way.

  One zone, clearly expressed in granite, charnokite, and sometimes in labradorite, corresponds to pressure values ranging from atmospheric to 1 kbar, rarely, 2-3 kbar. As the pressure increases further (to 6-8 kbar) the same rocks exhibit another zone characterized by a slowing down of the

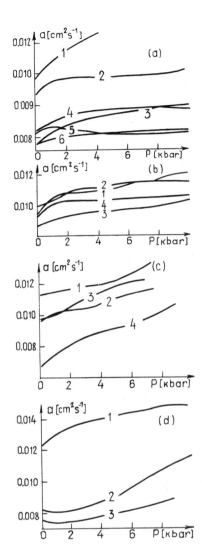

Fig. 3. Temperature conductivity versus pressure for some rocks of the Ukrainian Shield. a: 1,2 - equigranular granite partly cataclized, 3 - porphyry-like granite, 4,5,6 - coarse-grained labradorite; b: 1,4 - equigranular charnokite; c: 1,3 - pyroxenite, 2 - eclogite-like rock, 4 - diorite; d: 1,3 - biotite plagioclase gneiss (studied in three orthogonal directions).

increase of the temperature conductivity coefficient. Then, starting from 6-8 kbar, the coefficient a begins to grow again and a third zone is formed. It is worth noting that in labradorite the temperature conductivity increases as the pressure rises to 10 kbar.

In diorite, eclogite-like rocks and pyroxenite (Fig. 3c), the temperature conductivity coefficient behaves quite differently. In these rocks the coefficient of temperature conductivity grows monotonically with pressure; small flattened areas occur in the curves for diorite and eclogite-like rocks.

So far, a unique explanation of the origin of these zones, of their different widths or of their complete absence has been impossible. One may only suggest that the first zone with rather high gradients of the temperature conductivity growth is due to the improvement of the contacts between the constituent minerals, the decrease of the initial porosity and jointing of the rocks. This might lead to a decrease of thermal resistivity of the sample and, hence, to an increase of the temperature conductivity.

The second zone is the region where the increase in the temperature conductivity coefficient stabilizes as pressure rises. In this zone, the rock compaction ceases, whereas the contacts between the mineral grains remain sufficiently high. However, no structural change is yet observed here. Furthermore, in this zone, areas of considerable stress at the boundaries of separate grains may arise. This may lead to an increase in the phonon dissipation and, possibly, to a general slowing down of the temperature conductivity coefficient increase. The width of the zone correlates with the basicity of the rock as well. In acidic rocks it is relatively large (up to 2-7 kbar). In diorite and eclogite-like rocks the zone narrows. In pyroxenite it does not exist at all.

A further increase of pressure to 10 kbar evidently favours a partial change in the structural-textural properties of the mineral matter and a release of the local strain at the grain boundaries. This is responsible for the improvement of the thermal contacts and, hence, for the increase of the temperature conductivity coefficient in the third zone.

Interesting results were obtained by studying a sample of biotite plagioclase gneiss (Fig. 3, d). The measurements were made in three orthogonal directions. Under atmospheric pressure, isotropy of the temperature conductivity coefficient is observed. With increasing pressure, the a-values change differently along each of the three axes. Hence, there is a general increase of the temperature conductivity along one axis, whereas along the other axes the parameter shows only a slight increase with pressure in a low pressure range. However, pressure range from 5 kbar and more, the temperature conductivity increases rapidly with a considerable decrease in the anisotropy.

Conclusions

Experimental studies of the temperature conductivity coefficient variations in different rocks under high pressure involve great difficulties of methodical and technical nature. The results obtained in this work do not provide a unique description of the features of the thermal energy transfer in lithospheric minerals. At present, it can only be concluded that the temperature conductivity variation in the Ukrainian Shield rocks under high pressure is a complex function of many factors, including mineral composition, texture-structural features, elastic parameters, porosity, humidity, etc. The temperature conductivity of rocks is also strongly dependent on thermal contacts between individual grains of constituent minerals, their size and thermal properties. The mechanism of heat transfer in rocks and minerals under varying thermobaric conditions can be comprehensively studied only through the joint analysis of the thermal characteristics of the matter and their correlation to microtexture and other physical parameters.

References

Lebedev, T.S., 1975: Physical properties of lithospheric rocks under conditions prevailing at great depth. - In: The problems of physics of the earth in the Ukraine. - "Naukova Dumka", Kiev, 98-117. (Russ., Engl. abstract).

Lebedev, T.S., Korchin, V.A. & Vasilyaka, V.T., 1976: Hydrostatic investigations of elastic properties of mineral matter with high thermodynamic conditions. - Geofiz. Sb., Kiev, No 69: 26-34. (Russ., Engl. abstract)

Lebedev, T.S., Shapoval, V.I. & Pravdivy, A.A., 1979: Thermophysical measurements on minerals under high pressure and temperature. - Geofiz. Zhurn., No 2, Kiev, 17-26. (Russ., Engl. abstract)

Lyubimova, E.A., Starikova, G.N. & Shushpanov, A.P., 1964: Thermophysical investigations of rocks. - In: Geothermal studies. - "Nauka", Moscow, 115-174. (Russ.)

Magnitsky, V.A., Petrunin, G.I. & Yurchak, R.P., 1971: Temperature conductivity behaviour in some feldspars and plagioclase at 300-1200 K. - Dokl. AN SSSR, 199: 1058-1060. (Russ.)

Seipold, U. & Gutzeit, W., 1977: Temperature conductivity of rocks under high pressure. - In: Investigations of physical properties of mineral matter of the earth under high thermodynamic conditions. - "Naukova Dumka", Kiev, 154-160. (Russ.)

Shapoval, V.I., Seipold, U., Burtny, P.A. & Pravdivy, A.A., 1978: Temperature conductivity of some rocks of the Ukrainian Shield under high hydrostatic pressure. - In: Physical properties of rocks under high thermodynamic conditions. - "Elm", Baku, 250-251. (Russ., Engl. abstract)

Shapoval, V.I., Lebedev, T.S., Pravdivy, A.A. & Vasilyaka, V.T., 1978a: Some problems of the high pressure studies of thermophysical properties of rocks. - In: Physical properties of rocks under high thermodynamic conditions. - "Elm", Baku, 252. (Russ., Engl. abstract)

Shapoval, V.I. & Pravdivy, A.A., 1978: High pressure technique for measuring thermophysical characteristics of rocks. - In: Geophysical studies of the lithosphere in the Ukraine. - "Naukova Dumka", Kiev, 155-157. (Russ.)

Stiller, G., Seipold, U., Faber, J. & Folstadt, G., 1972: Heat conductivity studies under extremal conditions. - Geofiz. Sborn., Kiev, No 47: 13-16. (Russ., Engl. abstract)

Yurchak, R.P., Tkach, G.F., Petrunin, G.I. & Mahmud Mebed, 1973: Investigation of thermophysical properties of dielectrics at high temperature. - In: Thermophysical properties of solids. - "Nauka", Moscow, 83-87. (Russ.)

Numerical calculation of strong temperature gradients at interfaces between horizontally stratified sediments *

Günther Schroth

with 6 figures

Schroth, G., 1982: Numerical calculation of strong temperature gradients at interfaces between horizontally stratified sediments. – Geothermics and geothermal energy, eds. V. Čermák & R. Haenel, E. Schweizerbart'sche Verlagsbuchhandlung, Stuttgart: 141-147.

Abstract: In a cooling borehole, which has been previously heated, the so-called strong temperature gradients may occur at the interfaces of sediment layers. To show their dependence on the heating time as well as on the contrast between the thermal diffusivities, the temperature field around a borehole is calculated. A rapid increase and decrease of the strong temperature gradients corresponds to a short heating time while a longer heating time leads to an increase and longer persistence of the strong gradients. In particular, it is shown that strong *negative* gradients are induced at interfaces, where a sediment with high thermal diffusivity is covered by a sediment with low thermal diffusivity.

Author's address: Geophysikalisches Institut der Universität Karlsruhe, Hertzstr. 16, D-7500 Karlsruhe, Fed. Rep. of Germany

Introduction

Heat from hot oil or water flowing through a vertical borehole, which penetrates horizontally stratified sediments, will propagate differently into each sediment layer because of the varying thermal diffusivity. Consequently, strong temperature gradients result at the interfaces between the layers. In order to estimate the amount by which the temperature gradients increase, the 2-dimensional temperature field was calculated for cylindrical conditions around a borehole (Carslaw & Jaeger 1969).

Method

The calculations are executed with the Explicit Finite Difference method (Frank & Mises 1961, Marsal 1976) and are based on the following assumptions:
  - the borehole is vertical with respect to the sedimentary layers, and there are no lateral inhomogeneities;
  - the stationary temperature field of the stratified medium is used as the initial condition;
  - the vertical gradient is calculated at the centre of the borehole;
  - the influence of the casing and the thermal resistance of the contact

---

* Contribution No. 235, Geophysical Institute, University Fridericiana of Karlsruhe

surfaces are not taken into consideration;
- each layer has a sharp contrast of its thermal properties to the next layer;
- the boundary condition in time is the unit step function.

Results

A survey of a part of the temperature field is presented with the use of vertical and radial cross sections (Figs. 1 and 2). Two effects which influence the increase and decrease of the strong temperature gradients, as well as their maximum and the time when it appears, have been found. The existence of a strong gradient depends on both the heating time and the sedimentary layers. For two different heating times the time dependence of the strong temperature gradient is shown in Fig. 3. In Figs. 4a to d the maximum values of the strong temperature gradients and the time when the maxima occur are shown for different heating times and contrasts of the thermal properties. The results for a simulated production process from a completed borehole are shown in Fig. 5 and Figs. 6a to c. One can observe that the layers with a high diffusivity have a slightly flatter radial temperature distribution than the layers with a low thermal diffusivity. Nevertheless they cool down faster near the borehole. Hence a temperature decrease with depth inside the borehole is possible if a layer with low thermal diffusivity lies over a layer with high thermal diffusivity.

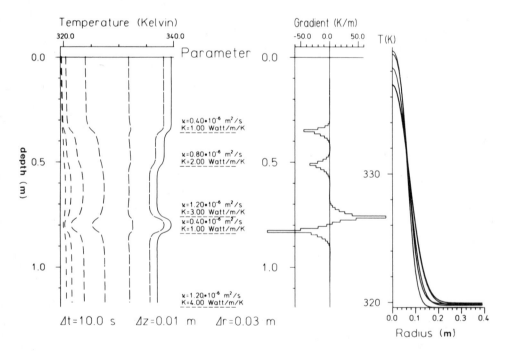

Fig. 1. A part of the temperature field with a depth of one metre and five different layers. The vertical temperature distributions at $r = n \cdot \Delta r$ ($n = 0$ to 6), the parameters, the temperature gradient, and five radial temperature distributions taken in the middle of each layer. Interaction of two strong gradients is established in the fourth layer.
Δt, Δz, Δr = discretisation steps for the Explicit Finite Differences.

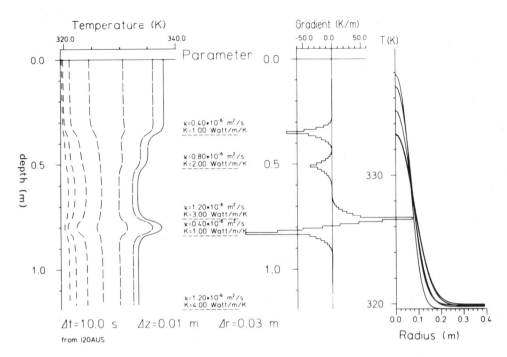

Fig. 2. Some time later the strong gradients have increased and inter-action of two strong gradients is also established in the second layer, i.e. the temperature gradient is no longer nearly parallel to the depth axis in the middle of the layer.

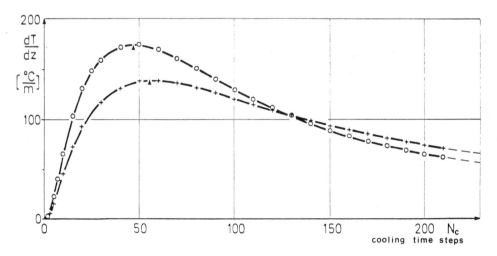

Fig. 3. The growth and disappearance of the strong gradient at an inter-face with a contrast of the thermal diffusivities and thermal conductivi-ties of 3 : 1 for two different heating times. The shorter heating time (30 time steps) causes an earlier and bigger maximum (shown by the symbol: ▲) of the strong gradient. For the longer heating times (150 time steps) the strong gradients remain large for a longer time.
+ = 150 time steps; o = 30 time steps.

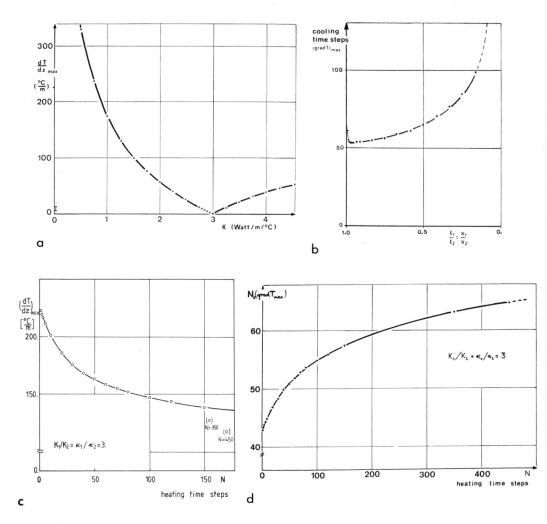

Fig. 4a-d. The maximum value of the strong gradient and its time of occurrence are both functions of the heating time and the contrast of the thermal diffusivities. Fig. 4a and 4b show the maximum and its time of occurrence for varying diffusivity contrasts and for a constant heating time. Similarly, Fig. 4c and d show the maxima and their times of occurrence for a constant contrast of 3 : 1 and for varying heating time steps.

Conclusions

The median gradients reach values of 0.5 °C/m over a distance of 5 m caused by a temperature difference of 20 °C and a thermal diffusivity contrast of 1 : 2. These gradients are greater than the so-called critical temperature gradient (Sammel 1968) which causes natural convection within the borehole. In the case of negative gradients the flow direction is contrary to the normal convection. Therefore, it follows that the strong gradients can be observed only in segments of the borehole where both the normal and induced convection are prevented.

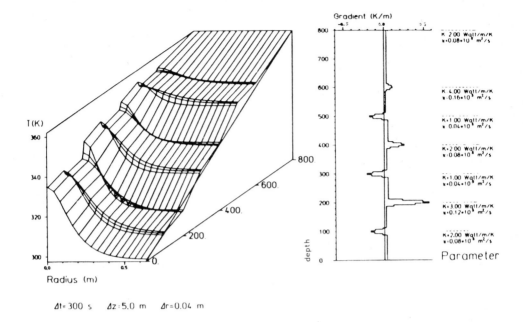

$\Delta t = 300$ s $\Delta z = 5.0$ m $\Delta r = 0.04$ m

Fig. 5. The temperature field after a heating time of 12 hours and a subsequent cooling time of 45 minutes. At this moment the temperature
gradients at the interfaces are the maximal strong gradients.

Acknowledgements. The research was carried out with financial support from
the European Communities under contract 321/78/2 EGD to the University of
Karlsruhe. The author is grateful to K. Fuchs for helpful discussions and
wishes to thank all members of the Geophysical Institute and the Computer
Centre of the University of Karlsruhe, where the numerical calculations
were made. Particular thanks go to S. Raikes, who corrected the English
translation, and Mrs. Di Pillo, who typed the manuscript.

References

Carslaw, H.S. & Jaeger, J.C., 1969: Conduction of heat in solids. - Oxford
    University Press.
Frank, PH. & v. Mises, R., 1961: Die Differential- und Integralgleichungen
    der Mechanik und Physik, II. - Dover Publication, Inc.
Marsal, D., 1976: Die numerische Lösung partieller Differentialgleichungen
    in Wissenschaft und Technik. - B.I. Wissenschaftsverlag AG, Zürich.
Sammel, E.A., 1968: Convective flow and its effect on temperature logging
in small-diameter wells. - Geophysics, 33, No. 6: 1004-1012.

Fig. 6 see page 146 and 147.

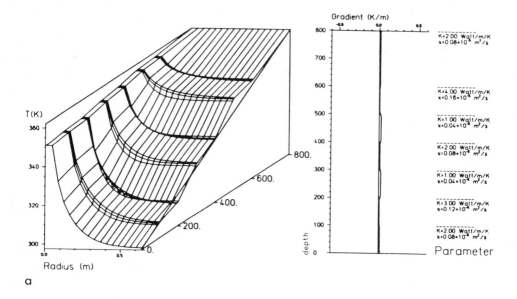

a

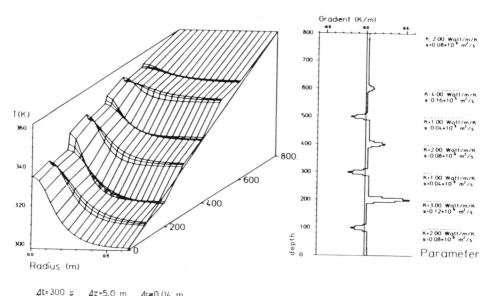

$\Delta t = 300$ s $\Delta z = 5.0$ m $\Delta r = 0.04$ m

b

Fig. 6a–d. In Fig. 6a the temperature field after a heating time of 12 hours is shown. The initial condition was the stationary temperature field of the stratified medium as seen at a radius r = 0.6 m. The cooling process begins at this time and generates the temperature fields of

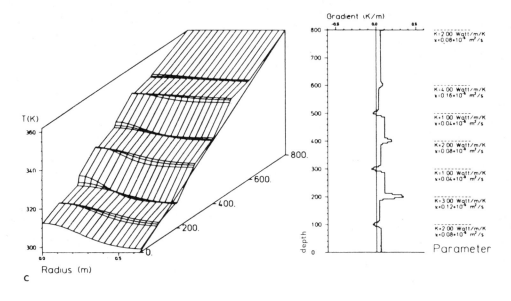

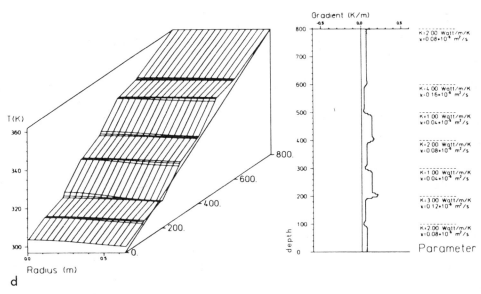

Figs. 6b – 6c. Fig. 6b is the same as Fig. 5. Fig. 6c shows the temperature field 12 hours and Fig. 6c 24 hours after the simulated production
process has been stopped.

Mathematical models of geothermal fields in gradient media

Simeon Kostyanev

with 1 figure

Kostyanev, S., 1982: Mathematical models of geothermal fields in gradient media. - Geothermics and geothermal energy, eds. V. Čermák & R. Haenel, E. Schweizerbart'sche Verlagsbuchhandlung, Stuttgart: 149-153.

Abstract: A mathematical model of the geothermal field in gradient media, i.e. in media with an arbitrary distribution of the heat transfer coefficient λ (z), is presented. In addition a certain volume V_O, where the heat transfer coefficient is different from the heat transfer coefficient in the surrounding media, is present. This model may represent the heterogeneous conditions in the surface and near-surface layer of the earth's crust.

Author's address: Higher Institute of Mining and Geology, Sofia-1156, Bulgaria

Introduction

Geothermal observations can be used for detecting heterogeneities of media in sedimentary beds or in the crystalline basement. The investigation comprises structural-tectonic elements, heterogeneities of hydrogeological nature, various inclusions (ore and intrusive bodies), etc. The above-mentioned heterogeneities cause the disturbances of the deep heat flow field on the earth's surface. By using the observation and analysis of the thermal regime in the surface layer, we can separate an anomalous component from the thermal field studied; this provides information on the subject under investigation. The basic interpretation of the measured anomalous thermal field consists in comparing the surface heat flow distribution with that calculated for some models. This procedure requires effective methods of calculation for direct problems of stationary thermal field distributions in heterogeneous media.

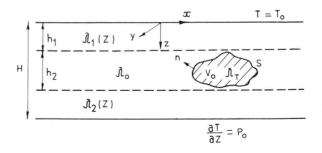

Fig. 1. Layer of arbitrary distribution of the heat transfer coefficient λ(z) containing heterogeneity V_O with a constant heat transfer coefficient.

Formulation of the problem

Suppose that there is a heterogeneity V_o with a constant heat transfer
coefficient in the bed $0<z<H$ with arbitrary distribution of heat transfer
coefficient $\lambda(z)$ (Fig. 1). It is assumed that the heterogeneity V_o is
located in a layer of constant $\lambda(z) = \lambda_o$, i.e.:

$$\lambda(z) = \begin{cases} \lambda_1(z), & 0<z<h_1 \\ \lambda_o, & h_1<z<h_1+h_2 \\ \lambda_2(z), & h_1+h_2<z<H \end{cases}$$

$$\lambda(M) = \begin{cases} \lambda(z) & \text{at } M \in V_o \\ \lambda_T & \text{at } M \in V_o \end{cases}$$

The temperature field in the layer $T(x,y,z)$ is to be found, as well as
the heat flow at the earth's surface (more exactly $(\partial T/\partial z)_{z=o} = P(x,y)$)
if constant heat flow is given at the lower surface of the layer ($z = H$),
and constant temperature at the upper surface ($z = 0$). The mathematical
problem is reduced to finding the function $T(x,y,z)$, satisfying the
equation:

$$\text{div}(\lambda(M) \text{ grad } T) = 0 \tag{1}$$

within the layer.
With conjugation conditions on the surface of disruption:

$$[T] = 0; \quad \left[\lambda \frac{\partial T}{\partial \overline{n}}\right] = 0, \tag{2}$$

where $\overline{n}$ is the normal to the disruption surface, and square brackets are
used to indicate the difference between the boundary values at both sides
of the disruption surface $\lambda(M)$. The following boundary conditions must
also be fulfilled:

$$T(x,y,z = 0) = \text{const}; \quad \left.\frac{\partial T}{\partial z}\right|_{z = H} P_o = \text{const} \tag{3}$$

with infinity condition:

$$\frac{\partial T}{\partial r} \to 0 \qquad \text{at } r = \sqrt{x^2+y^2} \to \infty. \tag{4}$$

The given problem can be solved by using the integral equation method, the
differential method and/or Galerkin type projection methods. The present
paper deals with the application of the integral equation method (Dmitriev
1969).

Calculation of the normal field

The general temperature field $T(x,y,z)$ is subdivided into a normal field
$T^{(n)}(z)$ (field without heterogeneities) and an anomalous field $T^{(a)}(x,y,z)$,
resulting from the disturbing effect of a heterogeneous body. In this way:

$$T(x,y,z) = T^{(n)}(z) + T^{(a)}(x,y,z) \tag{5}$$

The normal field is obtained by the solution of the problem:

$$\begin{cases} \dfrac{d}{dz}\left(\lambda(z)\,\dfrac{dT^{(n)}}{dz}\right) = 0 \\[2mm] \left[T^{(n)}\right] = 0; \ \left[\lambda(z)\,\dfrac{dT^{(n)}}{dz}\right] = 0 \\[2mm] T^{(n)}(z=o) = T_0; \ \dfrac{dT^{(n)}}{dz}\bigg|_{z=H} = P_o \end{cases}$$

The solution of this problem can be written in the following analytical form:

$$T^{(n)}(z) = T_o + P_o\,\lambda(H)\int_0^z \frac{d\lambda}{\lambda(z)} \tag{6}$$

Calculation of the anomalous field

The following problem is given for the anomalous field:

$$\text{div}\,(\lambda(z)\,\text{grad}\,T^{(a)}(x,y,z)) = 0; \ 0<z<H; \ -\infty<x,y<\infty \tag{7}$$

At the disruption $\lambda(z)$ conjugation conditions are satisfied:

$$\left[T^{(a)}\right] = 0; \ \left[\lambda(z)\,\frac{\partial T^{(a)}}{\partial z}\right] = 0 \tag{8}$$

At the surface of heterogeneity S the following conditions are satisfied:

$$\left[T^{(a)}\right]_S = 0; \ \lambda_o\left(\frac{\partial T^{(a)}}{\partial n}\right)_{outer} - \lambda_T\left(\frac{\partial T^{(a)}}{\partial n}\right)_{inner} = (\lambda_T-\lambda_o)\frac{\partial T^{(n)}}{\partial n}\bigg|_S \tag{9}$$

where $\left(\dfrac{\partial T^{(a)}}{\partial n}\right)_{outer}$ and $\left(\dfrac{\partial T^{(a)}}{\partial n}\right)_{inner}$ are boundary values of the

normal derivative from the outer and inner side of the surface S.
    The following homogeneous boundary conditions are satisfied at the layer surfaces:

$$T^{(a)}(x,y,z = 0) = 0; \ \frac{\partial T}{\partial z}\bigg|_{z=H} = 0 \tag{10}$$

and for infinity, the condition can be expressed:

$$\frac{\partial T^{(a)}}{\partial r} \to 0 \quad \text{at} \quad r = \sqrt{x^2+y^2} \to \infty. \tag{11}$$

It should be noted that in equation (7) $\lambda(z)$ is written instead of $\lambda(M)$, because $\lambda(M)$ differs from $\lambda(z)$ only in the region V_o, where $\lambda(z) = \lambda_o =$ const and $\lambda(M) = \lambda_T =$ const and for constant λ, equation (1) becomes $\Delta T = 0$ and does not depend on the λ-value.
    The boundary value problem (7-11) thus obtained for the anomalous field can be easily reduced to an integral equation. Suppose $G(M, M_o)$ is the Green's function for the boundary value problem:

$$\begin{cases} \text{div } (\lambda(z) \text{ grad } G (M,M_o)) = - \delta (r_{MM_o}) \\[2mm] [G] = 0; \quad \left[\lambda(z) \dfrac{\partial G}{\partial z}\right] = 0 \\[2mm] G\Big|_{z=o} = 0; \quad \dfrac{\partial G}{\partial z}\Big|_{z=H} = 0 \\[2mm] \dfrac{\partial G}{\partial r} \to 0 \text{ at } r \to \infty \end{cases} \tag{12}$$

Then the anomalous temperature field can be represented as a simple layer potential:

$$T^{(a)} (M) = \oiint_S \mu(M_o) \, G (M,M_o) \, d S_{M_o} . \tag{13}$$

The field represented in this way satisfies all the conditions of problem (8-11), except for those of normal derivative disruption on the surface S. Substituting the expression (13) into this condition yields the following integral equation:

$$\frac{\lambda_o + \lambda_r}{2} \mu (M) + (\lambda_o + \lambda_r) \oiint_S \mu (M_o) \frac{\partial G(M,M_o)}{\partial n_M} dS_{M_o} = \tag{14}$$

$$= (\lambda_T - \lambda_o) \frac{\partial T^{(n)}}{\partial n_M}$$

Having solved the integral equation, according to (13), we define the field within the layer. For solving equation (14) it is necessary to know the Green's function $G(M,M_o)$. Next, the method of calculation $G(M,M_o)$ is analysed.

Calculation of Green's function

If the origin of coordinates is transferred to the point $(x_o, y_o, 0)$, then the Green's function $G(M,M_o)$, where $M = (x,y,z)$, $M_o = (x_o, y_o, z_o)$, will become an axially symmetric function, satisfying the equation:

$$\frac{\lambda(z)}{r} \frac{\partial}{\partial r} \left(r \frac{\partial G}{\partial r}\right) + \frac{\partial}{\partial z} \left(\lambda(z) \frac{\partial G}{\partial z}\right) = - \frac{\delta(r)}{2\pi r} : \delta(z-z_o);$$

$$r = \sqrt{(x-x_o)^2 + (y-y_o)^2}$$

according to (12).

The solution of this equation can be given in the following form by means of Hankel's transform:

$$G(M,M_o) = \int_0^\infty I_o (mr) \, g (z,m) \, m dm, \tag{15}$$

where, according to (12), the function $g(z,m)$ satisfies the boundary value problem:

$$\frac{d}{d\lambda} \left(\lambda(z) \frac{dg}{dz}\right) - m^2 \lambda(z) g = - \frac{1}{2\pi} \delta(z-z_o), \tag{16}$$

$$[g] = 0; \quad \left[\lambda(z) \frac{dg}{dz} \right] = 0, \tag{17}$$

$$g \bigg|_{z=o} = 0; \quad \frac{dg}{dz} \bigg|_{z=H} = 0 \tag{18}$$

It is most convenient to represent the solution of the boundary value problem (16-18) in the following way:

$$g(z,m) = \begin{cases} g(z_o) \exp \left[- \int_z^{z_o} \frac{dz}{\lambda(z)y(z)} \right] & \text{at } 0 \le z \le z_o \\[4mm] g(z_o) \exp \left[\int_{z_o}^H \frac{z(z)dz}{\lambda(z)} \right] & \text{at } z_o \le z \le H \end{cases} \tag{19}$$

The new unknown functions $Y(z)$ and $Z(z)$ are connected with $g/m,m)$ by the relationships:

$$Y(z) = \frac{g(z,m)}{\lambda(z)g'(z,m)} ; \qquad Z(z) = \frac{\lambda(z)g'(z,m)}{g/z,m)}$$

and satisfy Riccati's equations:

$$Y'(z) + m^2\lambda(z)Y^2(z) = 1/\lambda(z) \qquad 0<z<z_o \tag{20}$$

$$Z'(z) + \frac{1}{\lambda(z)} Z^2(z) = m^2\lambda(z) \qquad z_o<z<H$$

under the initial conditions:

$$Y(z = 0) = 0; \quad Z(z = H) = 0 \tag{21}$$

After the solution of the initial value problem (20), $Y(z)$ and $Z(z)$ are found and, according to (19), the function $g(z,m)$ is therefore defined except for the constant factor $g(z_0)$. This quantity can be found from the conjugation condition at the point z_o:

$$\lambda (z_o-0) \frac{dg}{dz} \bigg|_{z_o-0} - \lambda (z_o+0) \frac{dg}{dz} \bigg|_{z_o+0} = \frac{1}{2\pi} \tag{22}$$

Substituting the equation:

$$\lambda (z_o-0) \frac{dg}{dz} \bigg|_{z_o-0} = \frac{g(z_o)}{Y(z_o)} ; \quad \lambda (z_o+0) \frac{dg}{dz} \bigg|_{z_o+0} = g(z_o)Z(z_o)$$

in (22), and
we define

$$g(z_o) = \frac{Y(z_o)}{2\pi(1-Y(z_o)Z(z_o))} . \tag{23}$$

In this way the Green's function $G (M,M_o)$ is completely determined.

References

Dmitriev, V, 1969: Electromagnetic fields in gradient media. - Trudy Moskovskogo gosudarstvenogo universiteta, Moscow, 130 pp. (Russ.)

Palaeogeothermal model for the Saxonian Erzgebirge*

Chr. Oelsner

with 6 figures and 1 table

Oelsner, Chr., 1982: Palaeogeothermal model for the Saxonian Erzgebirge. - Geothermics and geothermal energy, eds. V. Čermák & R.
Haenel, E. Schweizerbart'sche Verlagsbuchhandlung, Stuttgart: 155-
162.

Abstract: The complex of the Saxonian Erzgebirge is connected with
an anomaly of the terrestrial heat flow. The maximum values are
higher by more than 30 mW m^{-2} than the mean value of 67 mW m^{-2} for
the whole of the GDR territory. The heat flow at the Moho is as
high as 50 % of the surface heat flow. The temperatures of regional
metamorphism are too high in comparison with the most probable
value for the palaeocover. It is supposed that they were influenced
by an additional heat source, which was caused by a Precambrian
subduction process in the area of the SE-part of the Zentral-
sächsisches Lineament and its extension to the SE. The course of
the plate boundary is characterised by the gravity highs of Magdeburg, the Lausitz and the block of the Vorsudeten Mountains.
Quantitative palaeogeothermal data for the time from the Lower
Carboniferous to Upper Permian were obtained from coalification data.

Author's address: Sektion Geowissenschaften, Bergakademie Freiberg,
DDR-9200 Freiberg, German Democratic Republic

Introduction

In constructing palaeogeothermal models, geophysicists need geological
information to obtain the most probable solution.
    Such geological information includes
- the thermal gradient of regional metamorphism,
- the temperatures of neomineralization, and
- the results of age determinations.
    With the help of such data we developed a qualitative model of the
evolution of the Saxonian Erzgebirge.

Geothermal situation and rock metamorphism

The erzgebirge region is evidently a zone of high heat flow; see Heat
Flow Map of Europe compiled by Čermák & Hurtig (1979). This region is
characterised by a strong increase of heat flow from its northwestern,
northeastern and southern fronts with the adjacent territories (Oelsner &
Hurtig 1979). While the mean heat flow over the territory of the GDR is
67 mW m^{-2}, the mean value of the heat flow in the Erzgebirge amounts to
80 mW m^{-2}.
    On the Moho heat flow map of Central Europe (Oelsner 1978) the Erzge-

* Publication Nr. 1252, Sektion Geowissenschaften der Bergakademie Freiberg

birge also stands out with a high of more than 30 mW m^{-2}.

The thermal gradients of the regional metamorphism in the Erzgebirge determined from the temperatures of recrystallization are in the range of 55 - 70 K/km (Hofmann et al. 1979). This value seems to be too high with respect to the most probable value for the palaeocover. Therefore, we look for a relation between the high thermal gradients of the regional metamorphism and the high heat flow (Moho and surface).

Such high Moho heat flow must be connected with the processes of heat convection in the upper mantle.

We shall try to show a possible connection between the results of the age determinations on the metamorphic rocks from the eastern and western parts of the Erzgebirge and the deep seated geothermal processes.

K/Ar age data for rocks of the Erzgebirge show an age difference of 120 Ma between samples from the eastern and western part at a distance of 120 km (Hofmann et al. 1979). The age increases from west to east. From this fact we can deduce that the regional metamorphism advanced to thrust 0.1 cm/a in the western direction. If we suppose that the deep-seated geothermal processes have taken part in the generation of the regional metamorphism, the "age gradient" of 0.1 cm/a can be considered as the velocity of a metamorphism-generating front.

If we suppose the existence of a convection cell with a thickness of 100 km, ratio of thickness to wavelength 0.5, thermal diffusivity 10^{-6} m^2/s, and velocity 0.1 cm/a, we then get a Rayleigh number of 780, which exceeds the critical Rayleigh number by 0.1 %. This means that convection could occur.

Development of a geodynamical model

Some years ago Birch (1975) gave numerical data on the effect of a hot spot on a moving oceanic crust. His results for the velocities of 0.01 and 0.1 cm/a are compiled in Fig. 1, which shows the zone of molten material

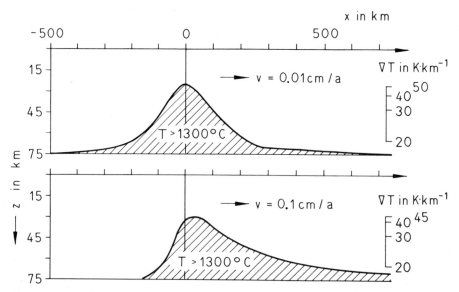

Fig. 1. Melting zone of a moving oceanic crust, caused by a hot spot. Redrawn after Birch (1975).

in the steady-state case. We completed Birch's figure with an additional
scale of temperature gradients. One can see, that the gradient above the
hot spot is 45 K/km, but that it already drops to only 27 K/km at a
distance of 200 km in the direction of the movement. That means that such
a hot spot gives rise to gradients which are in the range of the lower
limit for the thermal gradient of the regional metamorphism in the Erzgebirge. Hot spots are to be taken into consideration, for example in
connection with mantle diapirism.

   Stegena (1974) showed that the tectogenesis of all interarc basins
could be interpreted in terms of mantle diapirism. He gave some criteria
for the existence of interarc basins. Besides the high heat flow we state
only the existence of a positive gravity anomaly at the boundary range.
In view of this criterion it seems to be possible to interpret the Erzgebirge as a Precambrian ensialic basin.

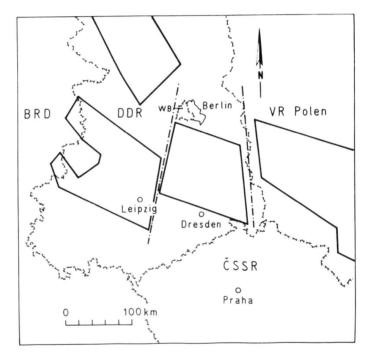

Fig. 2. Scheme of the positive gravity anomalies in a part of Central
Europe.

   Fig. 2 presents a scheme of the gravity anomalies in a part of Central
Europe. An arclike arrangement in the north of the Erzgebirge, the Iser
and Riesengebirge can be seen. The arc is formed from the west to the
east by the gravity highs of Magdeburg of the Lausitz and of the Vorsudeten. From the mutual position of the highs we can interpret a pole of
rotation which lies in the SE direction. The division lines can then be
interpreted as transform faults. They are depicted on the map of tectonic
faults in Europe by Grumbt et al. (1976). From the point of view of the
petrology of magma, such a model is supported by the results of Rösler &
Werner (1979). Werner (1980) discussed a plate tectonic model for the

thermal gradients which are separated from the effect of intrusive bodies on the one hand, and to get information on the thermal effects of these disturbing bodies, on the other hand.

Fig. 5 gives an example of the data from the Lugau-Oelsnitz area. Open circles represent the squared values of the measured reflection coefficients of the vitrinite. The intercept with the R^2-axis gives the value R_H of the reflection coefficient which is the result of the heating effect of an intrusive body. Full circles are the intrusion-corrected reflection coefficients. The slope is a measure of the palaeogeothermal gradient.

If we suppose that coefficients given by Buntebarth (1979) for the Oberrheintalgraben and by Buntebarth & Teichmüller (1979) for the intrusive of Bramsche are valid also for the Erzgebirge, the calculation of the palaeogeothermal gradient and of the "heating integral" yields the results compiled in Tab. 1.

Tab. 1. Palaeogeothermal results from coalification data.

Location	Palaeogeothermal gradient $^oC/km$	R_H	$\int Tdt$ 10^6 years
Lugau-Oelsnitz (L-Oe)	56.5	0.29	2.7
Zwickau (Zw)	40.5	0.61	5.7
Borna-Hainichen (Bo-Hai)	24.5	0.55	5.1
Döhlen-basin (Dö)	45.0	0.49	4.1

The undisturbed gradients are shown in Fig. 6. They belong to the period from the Lower Carboniferous to the Permian. It can be seen that the temperature gradients were the highest in the Westphalian D; this was the result of the highest rate of sedimentation.

The values of $\int Tdt$ from Tab. 1 can be used to test different intrusive models. The diagrams of Mundry (1968), for example, are useful. If we look at a sphere with radius r = 1 km, then the "heating integral" has the value

$0.4 \cdot 10^6$ (years · K) if the cover is 1000 m
$1.0 \cdot 10^6$ "   the cover is 800 m
$11.0 \cdot 10^6$ "   the cover is 500 m.

In such a manner we can seek some radius - cover relations for the values given in the last column of Tab. 1.

Conclusion

As next steps we have to transform this qualitative evolutionary model into a quantitative model and to combine the results with vitrinite data and other palaeotemperature data. By such means it should be possible to get an insight into the evolution of the heat flow in the area of the Erzgebirge from the Precambrian to the present.

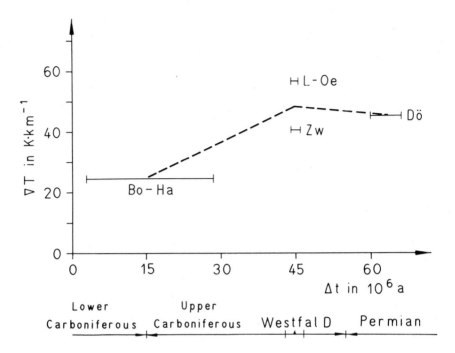

Fig. 6. Undisturbed coalification temperature gradients for the period of the Lower Carboniferous-Permian.
Abbreviations (see also Table 1):

Bo-Hai   Borna - Hainichen
L-Oe   Lugau - Oelsnitz
Zw   Zwickau
Dö   Döhlen basin

References

Birch, F.S., 1975: Conductive heat flow anomalies over a hot spot in a moving medium. - J. Geoph. Res. 80, 35: 4825-4827.
Buntebarth, G., 1979: Eine empirische Methode zur Berechnung des geothermischen Gradienten aus dem Inkohlungsgrad organischer Einlagerungen in Sedimentgesteinen mit Anwendung auf den mittleren Oberrheintalgraben. - Fortschr. Geol. Rhld. u. Westf. 27: 97-108.
Buntebarth, G. & Teichmüller, R., 1979: Zur Ermittlung der Paläotemperaturen im Dach des Bramscher Intrusivs aufgrund von Inkohlungsdaten. - Fortschr. Geol. Rhld. u. Westf. 27: 171-182.
Čermák, V. & Hurtig, E., 1970: Heat Flow Map of Europe. - In: V. Čermák & L. Rybach (Eds.), Terrestrial Heat Flow in Europe. - Springer-Verlag, Berlin, Heidelberg.
Chaloupský, J., 1980: The precambrian tectogenesis in the Bohemian Massiv. - Geol. Rundschau 67, 1: 72-90.
Grumbt, E. et al., 1976: Tektonische Bruchstörungen 1 : 6 000 000. - In: Jubitz, K.B. (Ed.), Materialien zum tektonischen Bau von Europa. - Veröff. Zentralinst. Phys. d. Erde, Nr. 47, Potsdam.
Hofmann, J. et al., 1979: Fazies und zeitliche Stellung der Regionalmetamorphose im Erzgebirgskristallin. - Z. Geol. Wiss. Berlin 7: 1091-1106.

Hvoždăra, M. & Rosa, K., 1979: Geodynamic effects of thermoelastic stresses due to a linear heat source. - In: Geodynamic Investigations in Czechoslovakia, Final Report. - VEDA, Bratislava: 53-63.

Karweil, J., 1955: Die Metamorphose der Kohlen vom Standpunkt der physikalischen Chemie. - Z. Deutsch. geol. Ges. 107: 132-139.

Künstner, E., 1974: Vergleichende Inkohlungsuntersuchungen unter besonderer Berücksichtigung mikrophotometrischer Reflexionsmessungen von Kohlen, Brandschiefern und kohlehaltigen Nebengestein. - Freiberger Forsch.-H. C 287, VEB Dt. Verl. f. Grundstoffindustrie, Leipzig, 120 S.

Mundry, F., 1968: Über die Abkühlung magmatischer Körper. - Geol. Jb. 85: 755-766.

Oelsner, Chr., 1978: Wärmestrom und Temperatur an der Moho in Mitteleuropa. - Rev. Roum. Geol., Geophys. et Geogr.-Geophysique 22: 49-58.

Oelsner, Chr. & Hurtig, E., 1979: Zur geothermischen Situation im Erzgebirge. - Freiberger Forsch.-H. C 350: 7-17.

Rösler, H.-J. & Werner, C.D., 1979: Petrologie und Geochemie der variszischen Geosynklinalmagmatite Mitteleuropas, Teil I. - Freiberger Forsch.-H. C 336, VEB Dt. Verl. f. Grundstoffindustrie, Leipzig, 160 S.

Stegena, L., 1974: Geothermics and tectogenesis in the Pannonian Basin. - Acta Geologica Academicae Scientiarum Hungaricae, T 18, 3-4: 217-266.

Weinlich, F.H., 1979: Spezielle Inkohlungsuntersuchungen an organischen Substanzen bzw. Kohlen an ausgewählten Proben aus dem Süden der DDR, (unpublished report) Sektion Geowissenschaften der BA Freiberg.

Werner, C.D., 1980: Proterozoische Metabasite im Sächsischen Grundgebirge (DDR). - Ophiolitmonograph of UK 2 of PK IX, Moscow, in press.

Comparative gravimetric and magnetometric investigations in and near the central graben (Ruhr-valley graben) in the province of Limburg (The Netherlands)

J. W. Bredewout

with 2 figures

Bredewout, J.W., 1982: Comparative gravimetric and magnetometric investigations in and near the central graben (Ruhr-valley graben) in the province of Limburg (The Netherlands). - Geothermics and geothermal energy, eds. V. Čermák & R. Haenel, E. Schweizerbart'sche Verlagsbuchhandlung, Stuttgart: 163-167.

Abstract: In order to examine the existence of a suspected magmatic intrusion underneath the Peel-boundary fault and the Ruhr-valley graben a gravimetric and magnetometric survey was started. Laboratory investigations on rock samples collected in a coal mine in the area as well as underground gravity observations carried out in that coal mine were used as additional information. A preliminary interpretation of the results seems to confirm the existence of an intrusion. This paper is a summary of the project. More details and a final conclusion will be given in a publication as soon as the work is completed.

Author's address: Vening Meinesz Laboratorium- IVAU, P. O. Box 80.021, 3508 TA Utrecht, The Netherlands

Introduction

The subsurface temperature map of The Netherlands does not exhibit striking temperature anomalies. For the northern part, where an enormous amount of data from oil and gas wells is available, this map is very reliable.

In the southern part of The Netherlands, however, useful data were available from only a few, and not even recent, deep boreholes; hence the map in that area is based mainly on data from shallow wells (depth less than 1500 m). Since the temperatures at shallow depths are strongly influenced by ground water flow, the absence of temperature anomalies on the map in this area does not necessarily exclude the possibility of temperature anomalies at greater depths.

In fact, in the survey area indicated in Fig. 1 we had three indications beforehand that a temperature anomaly (caused by a basic magmatic intrusion) might exist: (1) the temperature gradient in the Carboniferous at depths of 500–1000 m as observed in a few boreholes and in the German coal mine Sophia Jacoba at Hückelhoven is slightly higher than in the surrounding area (45 °C/km compared to 37 °C/km); (2) the degree of coalification of the coal from the Sophia Jacoba (Teichmüller 1971) and in the shaft of the unexploited Dutch mine Beatrix is relatively high; (3) there is a lack of correlation between the existing gravity and magnetic maps of the area.

The aim of the present research project is to establish the presence or absence of the suspected intrusion. For that purpose the following investigations are in progress: (1) a gravimetric and magnetometric survey at the surface; (2) laboratory measurements (density, susceptibility,

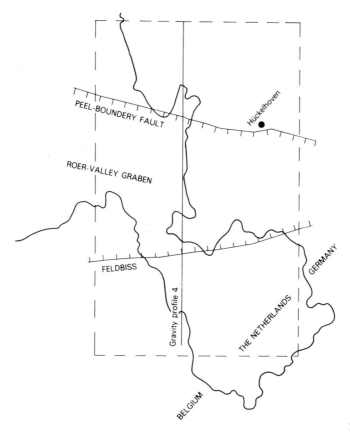

Fig. 1. Situation map.

palaeo-temperature) on rock samples from the Sophia Jacoba; (3) underground gravity observations in the Sophia Jacoba.

Gravimetric and magnetometric survey

The existing gravity map of the area (van Weelden 1957) correlates with
the known geological structures: a minimum of -25 mGal in the Ruhr-valley
graben and slightly positive values on the neighbouring horsts. The
magnetic map (Veldkamp 1951) does not show such a good correlation:
instead of running parallel to the major faults, the contour lines cross
the Ruhr-valley graben. Inspection of the adjacent part of the German
aeromagnetic map (Bundesanstalt 1976) revealed that the centre of a
positive anomaly (250 gamma) is located at Hückelhoven. The detailed
magnetometric survey we performed in the area confirmed the German map and
showed that this anomaly also extends over Dutch territory. As an
explanation for this anomaly the existence of a basic magmatic intrusion
lying between 4 and 15 km in depth, and having horizontal dimensions of
40 km (in NW-SE direction) by 30 km (in NE-SW direction), with a susceptibility of 2×10^{-3} SI units was already assumed by Bosum (1965).
    Inspection of two of the eight SW-NE running gravity profiles measured
showed that at a depth greater than of the deepest boreholes in the area

(1500 m), either at the SW-side rocks of relatively low density or at the NE-side rocks of relatively high density must be present. The latter interpretation agrees very well with Bosums assumption of a basic intrusion. As an example is depicted in Fig. 2 the observed gravity anomaly along profile 4. The full line is the anomaly calculated for a structural model based on the well known geology in the SW (left side) and the NE (right side). This calculated anomaly has been adapted to the observations at the far SW-end of the profile and coincides with the observations in the middle, while at the NE-end it deviates from the observed anomaly. This can be explained by assuming a structure of high density in the deep subsurface.

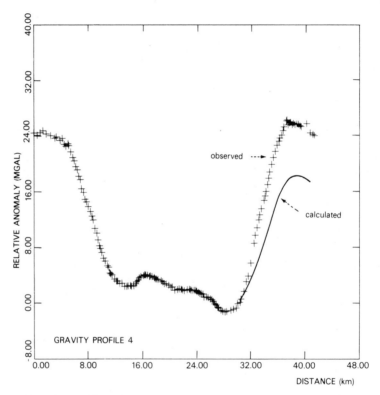

Fig. 2. Gravity anomalie of profile 4.

Laboratory measurements on rock samples

The magnetic susceptibility, the remanent magnetisation, and the density of rock samples from the coal mine Sophia Jacoba were measured. The average density was 2.68 g cm^{-3}. The susceptibilities were low (less than 10^{-3} SI-units) and no remanent magnetisation was found. Only a few samples from one location at a depth of 700 m showed deviating values for the density (3.1-3.5 g cm^{-3}) and for the susceptibility (3x10^{-3} SI-units). These rock samples turned out to be siderite and were lying as small concretions in a thin layer just underneath a coal seam. Their remanent magnetisation was negligible. Thus no rocks were found with such a high susceptibility or remanent magnetisation that they could (as an alternative for Bosums assumption) be the cause of the magnetic anomaly.

One of the siderite samples contained some quartz veins with small fluid inclusions. The homogenisation temperature of these inclusions appeared to be 204 °C. This means that the sample must once have been at a temperature of at least 204 °C, which is much higher than that correspond to the depth at which it was found or at which it could occur. This provides a further indication for an intrusive body at high temperature.

Underground gravity observations

From underground gravity measurements at various depths at stations lying vertically one below the other, one can calculate the densities of the rocks situated between the stations. This is true only if the vertical gradient of gravity can be described by the free-air correction plus the Bouguer correction. However, if a local mass deficiency or surplus is located somewhere near the stations, i.e. if the density is not only a function of the depth but also of the horizontal coordinates, then the magnitude of the vertical gradient of gravity is influenced by this and can no longer be used for density determinations.

Conversely, if the densities are known from laboratory measurements on rock samples, then by measuring the vertical gravity gradient one can determine whether at a greater depth an unknown structure with a differing density is present or not. For this reason we measured the vertical gradient at 5 locations in the Sophia Jacoba. From these data we deduced pseudodensities for the Carboniferous and for the overburden. For the Carboniferous we found 2.62, 2.71, 2.57, 2.69 and 2.68 g cm^{-3}, for the overburden 2.00, 2.02, 2.13, 1.97 and 2.07 g cm^{-3}. The maximum error is ± 0.04 g cm^{-3}. This is rather big because of the considerable corrections for topography, mine shafts and local faults. Some of the deduced pseudo-density values agree within the accuracy of the laboratory determinations (2.68 and 2.15 g cm^{-3}), but others deviate by more than the maximum error. Therefore the qualitative conclusion from these results is that the presence of a deep-lying structure of high density is not unlikely.

Conclusion

The assumption of a magmatic intrusion underneath the Peel-boundary fault and the Ruhr-valley graben was previously based on the following evidence: a high temperature gradient, a high degree of coalification and a magnetic anomaly. The existence of this intrusion is confirmed or at least not disproved by our investigations: the surface gravitational field, underground gravity observations, measurement of palaeo-temperature and susceptibility determinations.

Acknowledgement. This research is part of the Dutch National Research Programme for Geothermal Energy and is made possible by financial support from the Ministry of Economic Affairs. We are grateful to the management and staff of the Sophia Jacoba for enabling us to perform the underground measurements and to Prof. Schuiling c.s. for carrying out the analysis of the siderite sample.

References

Bosum, W., 1965: Interpretation magnetischer Anomalien durch dreidimensionale Modellkörper zur Klärung geologischer Probleme. - Geol. Jb. 83: 667-680.

Bundesanstalt für Geowissenschaften und Rohstoffe, 1976: Karte der Anomalien der Totalintensität des erdmagnetischen Feldes in der Bundesrepublik Deutschland.

Teichmüller, M. & R., 1971: Das Revier von Aachen-Erkelenz, b) Inkohlung. - Fortschr. Geol. Rheinl. u. Westf. 19: 69-72.

Veldkamp, J., 1951: The Geomagnetic Map of The Netherlands reduced to 1945.0. - Kon. Ned. Meteorol. Inst. publ. No. 134.

Van Weelden, A., 1957: History of gravity observations in The Netherlands. - Verh. KNGMG, Geol. ser. XVIII: 305-309.

Geothermal investigations in oil-bearing fields

T. Velinov and K. Bojadgieva

with 5 figures

Velinov, T. & Bojadgieva, K., 1982: Geothermal investigations in oil-bearing fields. - Geothermics and geothermal energy, eds. V. Čermák & R. Haenel, E. Schweizerbart'sche Verlagsbuchhandlung, Stuttgart: 169-174.

Abstract: A survey of studies related to the application of geothermal data obtained from oil and gas investigations is presented. Structural problems connected with oil and gas saturation were determined by temperature distribution. In accordance with the data on the geothermal gradient, the reservoir properties of large regions were evaluated. The distribution of clayey matter was closely studied since it controls the porosity and permeability of the layers. Some other tasks related to the exploration of hydrocarbon deposits were discussed, too.

Authors' address: Enterprise for Geophysical Exploration and Geological Mapping, Chr. Kabakchiev Str. 23, Sofia 1505, Bulgaria

Introduction

The geothermal surveys are widely used to solve a number of tasks in oil-gas geology. The main characteristic of the thermal fields is the heat flow (q). Some of the results of the geothermal surveys in the sediments of Northern Bulgaria are presented in this paper. The temperature (T) and the gradient (G) were used in the above surveys.

Regional investigations in northern Bulgaria

In studying the reservoir features of the deposits it is important to know the content of the clay component. Thus to determine the possibility of using the gradient values for the study of the clay content, we compared them with the indices of the gamma ray log. Carbonate, clayey-carbonate and marly deposits prevail in the sedimentary layers of the area explored. It is known that, because of the ^{40}K content in the clays, the gamma ray index is the higher, the higher the clay content is. The gradient values were compared with the parameter α, which is the ratio of the gamma index of a layer with unknown clay component ($I\gamma$) to the gamma index of a pure clay layer ($I\gamma^{ce}$). The comparison of α-values with the gradient calculated for layers occupying 500 m depth intervals is given in Fig. 1. It is evident that there is a relationship between α and G for the deposits studied; the values of the geothermal gradient can therefore be used for the study of the distribution of the clay-marl component in the sediments.

Maps of the gradient distribution in the intervals of 500-1000 m, 1000-1500 m, 1500-2000 m, 2000-2500 m, 2500-3000 m and 3000-3500 m have been drawn (Fig. 2).

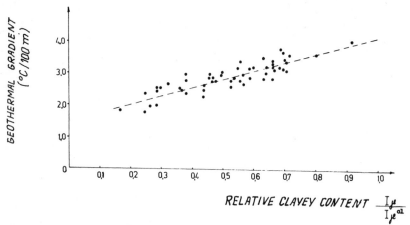

Fig. 1. Relation between the geothermal gradient (G) and the specific clay content $I\gamma/I\gamma^{ce}$ for the sediments in Northern Bulgaria.
$I\gamma$ - Indications of gamma ray log compared to surveyed layer,
$I\gamma^{ce}$ - Indications of gamma ray log compared to the clay layer.

Three characteristic intervals of gradient values have been distinguished: 1.5-2.5, 2.5-3.0 and above 3.0 °C/100 m. From the studies (Bojadgieva & Velinov 1977) it is evident that these intervals are characteristic of limestones and dolomites with no clay component, limestones and dolomites with more than 20 % of clay component and pure clayey-marley deposits, respectively. A joint analysis of these maps allows a study of the clay component distribution in horizontal and vertical directions.

It is evident from Fig. 2 that the most widespread distribution of clayey deposits is characteristic of the interval 1500-2000 m. It is built up mainly of deposits of the Lower Cretaceous and can be considered as a regional thermally insulating horizon and a reliable screen for the generation of oil-gas fields.

The hydrocarbon accumulations are related mainly to deposits characterized by gradient values from 2.5 to 3.0 °C/100 m.

Similar surveys were conducted on geological-geothermal profile lines (Fig. 3). The geothermal zones occuring on the vertical sections include gradient values analogous to those shown on the maps. The change of the lithology of the individual stratigraphical horizons could be traced in these zones. On Profile I a transition from clayey-marley deposits in the west to a pure carbonate section in the east is observed for the horizons of the Palaeogene and the Upper Cretaceous. On Profile II (parallel to Profile I and shifted to the east) the geothermal zones show unity of lithology in the stratigraphical horizons. The intervals of geothermal gradient facilitate the tracing of the character of sedimentation between the stratigraphical horizons.

Geothermal surveys for exploration of hydrocarbon accumulations

In view of the high cost of deep boreholes, we began to investigate the distribution of the temperature field in the oil-gas regions at relatively shallow depths. The first experiments were carried out in wells 200 m deep situated above the productive oil field. Thermometers with the accuracy of ± 0.1 °C were installed in all the wells and the temperature was

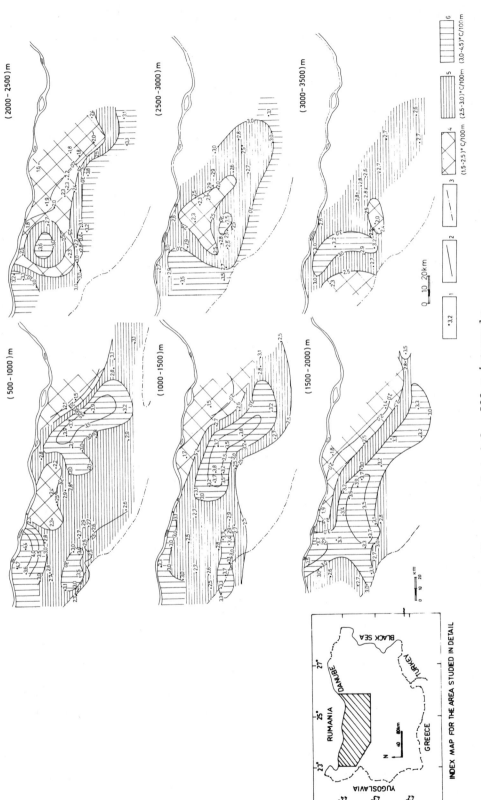

Fig. 2. Maps of the geothermal gradient calculated for 500 m intervals.
Explanations: 1. Position of the well and the value of the geothermal gradient; 2. Isolines of the geothermal gradient; 3. The boundary between the Moessian Plate and the Balkan Foreland zone; 4. (1.5–2.5); 5. (2.5–3.0); 6. (3.0–4.5) mean gradient values, in °C/100 m.

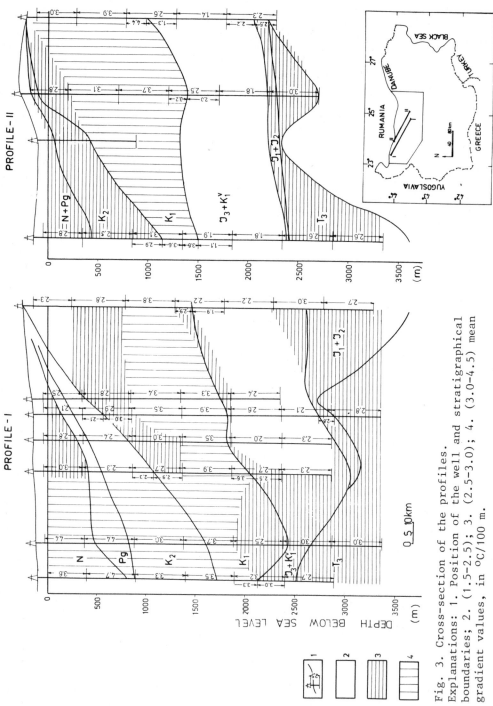

Fig. 3. Cross-section of the profiles.
Explanations: 1. Position of the well and stratigraphical boundaries; 2. (1.5–2.5); 3. (2.5–3.0); 4. (3.0–4.5) mean gradient values, in °C/100 m.

measured periodically at 200 m depth. In order to ensure the required
accuracy, a special calibration was performed on the thermometers used. It
turned out that the temperature at 200 m depth above an oil-gas-bearing
region of the field (wells W-1; W-42; W-44) was 2-3 °C higher than that in
the water bearing part (Fig. 4). This fact had already been established
(Sardarov & Suetnov 1975) but it was confirmed for the first time in
Bulgaria.

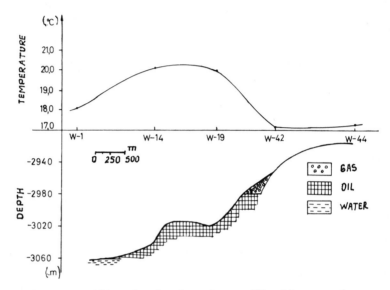

Fig. 4. Temperature change at 200 m depths for the profile line passing
over the oil-bearing field.

In connection with the positive results obtained from the study of the
temperature near the earth's surface it was resolved that surveys should
be carried out at another oil-gas field in wells 5 m deep. The geological
structure of this field is very complex, although an anticline structure
is established, the oil-bearing stratum is observed only at individual
sites. The field is probably of lithological type. During the surveys it
was suggested that the bed might be intersected by faults, which were not
recorded by the seismic surveys. On the basis of a complex analysis of
the drilling data and of the geophysical and geological data, Borissova
(1980) mapped two hypothetical faults with NE - SW directions (Fig. 5).
These faults may control the location of the oil-bearing layer, as well as
its saturation.

The near-surface temperatures above that field were measured at a
stationary thermal regime in wells 5 m deep located at intervals of 250 m
along four profile lines (Fig. 5). In order to avoid accidental errors in
the temperature curves, the temperature at each point is taken as an
arithmetical mean of three temperature values, one measured at a given
point and the other two at neighbouring points. The temperature curves of
the four profile lines are shown in Fig. 5. An increase of temperature by
up to 0.9 °C along profile I of the region, including shallow wells no. 4,
5 and 6 located between both faults, was observed. The 1 °C increase of
temperature in wells no. 4, 5 and 6 along profile III of the region is
analogous. Profile line II is located to the west of profile lines I and

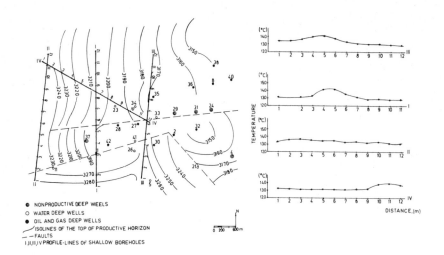

Fig. 5. Temperature measurements in boreholes 5 m deep at 250 m distance interval along four profile lines over oil-gas-bearing structure.

III in the nonproductive part of the field. A slight increase of the temperature (up to 0.3 °C) in wells no. 2, 3 and 4 is observed. That inexpressive temperature maximum is probably related to the reduced influence of the fault passing in the vicinity of the deep well no. 11, while the fault located to the north of it does not affect the temperature curve of profile line II due to its insignificant amplitude. The temperatures measured along profile IV increase by 0.7 °C in the region of the shallow wells no. 10, 11 and 12 located in the vicinity of the fault.

The results show the usefulness of the temperature measurements in the shallow wells for the study of the deep faulting in the oil-gas-bearing regions.

References

Bojadgieva, K. & Velinov, T., 1977: O svyazi teplovogo polya so structurno geologicheskim stroyeniem Bolgarii. - Proc. XXII, Internat. Geophys. Symp., Praha. (Russ.)

Borissova, J., 1980: Izutchavane zakonomernostite na izmenenie na fizitceskite svoistva na skalite v raiona mejdu rekite Iantra i Ogosta po sondajno-geofizitchni danni. - Geofond XG, Sofia. (Bulg.)

Sardarov, S.S. & Suetnov, V.V., 1975: Teplovoy potok formiruyemi neftegasonosnim structurami. - Sovetskaya geologya, No 2: 122-126. (Russ.)

A joint analysis of seismological and geothermal parameters

I. A. Kireev and N. V. Kondorskaya

with 2 figures and 2 tables

Kireev, I.A. & Kondorskaya, N.V., 1982: A joint analysis of seismological and geothermal parameters. - Geothermics and geothermal energy, eds. V. Čermák & R. Haenel, E. Schweizerbart'sche Verlagsbuchhandlung, Stuttgart: 175-180.

Abstract: The location of earthquake hypocentres and the Vp velocity structure of the earth's crust are used for comparison with deep geotherms of the sublatitudinal cross-section of the Caucasus. Geothermal models accounting for the behaviour of the thermal conductivity coefficient and the empirical relationship between the Vp velocity and the heat production in rocks are developed on the basis of laboratory data. Geotherms, mantle heat flow and thermal gradients are calculated for the profile Black Sea - Caspian Sea. Variations of the Moho temperature are in the range of 450-720 $^{\circ}$C, the heat flow from the upper mantle being 8-19 mW/m^2. The accuracy of calculating geotherms is estimated. The significance of the isothermal surface at 400 $^{\circ}$C is emphasized as bounding the depth of the majority of hypocentres. The correlation between the sedimentary layer thickness and the temperatures beneath is discussed.

Authors' address: Institute of Physics of the Earth, USSR Academy of Sciences, B.Gruzinskaya 10, Moscow, USSR

Introduction

Much attention has lately been paid to research on deep geothermal parameters in connection with the study of the structure of the earth's crust (Čermák 1979, Hurtig & Stromeyer 1979, Kutas 1979). An attempt to apply these parameters to the seismic regions of the Caucasus is made in this work.

In a high-temperature medium, stresses relax; thus no earthquakes are caused (Popov 1963). At a low temperature the viscosity of rocks is high and the release of stress may cause an earthquake (Artyushkov 1972). In this connection the P-T conditions of seismic volumes are significant. The comparison of such data for real profiles with laboratory results in this field may help improve the petrological model of the earth's crust.

Geophysical anomalies correspond to the peculiarities of the geological structure of the earth's interior (Kunin & Volvovsky 1976), which are supposed to be revealed in geothermal fields. An attempt is made to account for the seismic Vp velocity anomalies of the sublatitudinal deep-seismic-sounding profile of the Caucasus (Fig. 1) (Kondorskaya et al. 1980) in heat generation models of the earth's crust according to Rybach (1979).

The effect of the sedimentary layer on the thermal regime of underlying layers of the earth's crust is studied for the cross-section mentioned above.

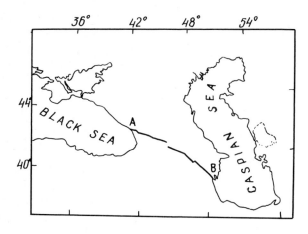

Fig. 1. The location of the
Caucasus DSS cross-section.
A - Black Sea coast,
B - Caspian Sea coast

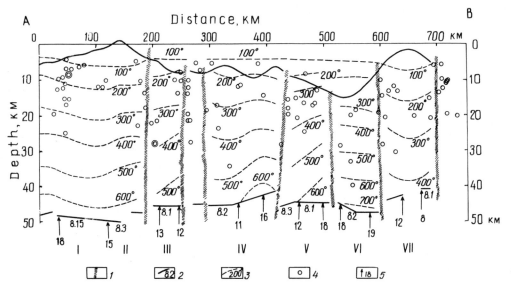

Fig. 2. The structure of the earth's crust along sublatitudinal DSS
cross-section of the Caucasus (interpreted by G. V. Krasnopevtseva).
1 - deep faults; 2 - boundaries of consolidated crust and Vp values;
3 - geotherms; 4 - projections of earthquake foci onto the figure plane;
5 - mantle heat flow values.

Seismological parameters

According to the velocity field along the above profile (Fig. 2), the
earth's crust of the intramountain basins of the Caucasus has a layer-
fractured structure (Kondorskaya et al. 1980). The units of the earth's
crust are considered as being bounded by fractures.
    The above profile was divided into seven units. The units of the
earth's crust in question are characterized as having earthquake foci. The
velocity pattern of the medium is designed for each unit (Kondorskaya et
al. 1980). The units under consideration are studied by both geological

and geophysical methods. Tectonics, the relief, the age and the type of
the crust are taken into account. There is some information available on
the results of drilling, heat flow measurements, the composition of sediments and crystalline rocks, etc.

The parameters of earthquakes for the units involved are taken from
the Catalogue of Large Earthquakes in the USSR (New Catalogue, 1977).
The epicentre coordinates are determined by use of both global and regional hodographs with an accuracy of 0.1° for the majority of events. The
depth of the hypocentre is determined by different methods, the resolving
capacity being as large as 5 km. Earthquakes of magnitude of 2.5 and higher
are considered. The precision of estimating the magnitude is up to 0.2
for the catalogue data.

Geothermal parameters

The geotherms of the units of the earth's crust are calculated in comparison with the seismological parameters concerned. A three-layered petrological model of the earth's crust was considered: (a) sedimentary layer,
(b) metamorphic rocks and "granites", (c) gabbro. The thermophysical
properties that are a continuous function of the pressure and temperature
within each layer were considered.

The steady-state heat conduction equation with a free term defining
heat generation is used. The constant temperature on the surface is used
as a boundary condition. For computation of geotherms the most carefully
studied points of the profile are chosen. Two geothermal models are constructed for the vertical cross-section of the crust at each of these
points i.e., heat conduction and heat generation models. Both geothermal
models are based on laboratory measurements of the parameters of the
assumed types of rocks that compose the unit under certain P-T conditions.
The empirical relations can be chosen to determine these models. Thus, the
relation between heat generation and Vp velocity in consolidated crust
down to 35 km (Rybach 1979) has been chosen:

$$A (Vp) = a \cdot exp (- b \cdot Vp) \qquad (1)$$

where A = heat generation, Vp = compressional wave velocity, a and b =
constants, characterising granite, diorite and gabbro under corresponding
P-T conditions. The thermal conductivity (K) of rocks depends inversely
on temperature (T):

$$K (T) = \frac{Ko}{1 + \alpha T} \qquad (2)$$

where Ko is the thermal conductivity at room temperature, and α is the
coefficient of the respective rock. Relation (2) is satisfactory for most
dry igneous rocks and sediments (Zharkov 1978). The precision of the
thermal profile obtained is determined by the accuracy of the measurements
of the values of A, K and Q (surface heat flow), as well as the assumed
limits of their variations. It should be noted that the accuracy of the
surface heat flow measurements is about 10 % (Kutas 1978). The precision
of the laboratory measurements of the thermal conductivity coefficient and
the content of radioactive elements in the rocks is about 5-10 %. The
variation of these values for different samples of the same type of rock
is approximately 20 % on the average. By taking all these errors into
account, one can evaluate an error in the temperature values for different
depths. According to our estimates this error does not exceed 60 °C at a
depth of 10 km. However, it increases with depth, reaching 150 °C at the

Moho discontinuity. In other words, this means that the displacement in depth of the isotherms in the upper layers of the earth's crust (up to 10 km) can reach 3 km, and at the bottom of the crust, 12 km.

The upper mantle heat flow Q_M is calculated by means of the equation

$$Q_M = Q - \int_0^{z_M} A\,(z)\,dz \tag{3}$$

where z_M is the depth of the Moho discontinuity.

The predominant composition of rocks of the surface layer was taken into account in designing the geothermal models for closely studied parts of the profile (Milanovsky 1968). The mean values of 1.25 $\mu W/m^3$ for heat production and 1.68 W/m.K for heat conductivity were accepted for sediments (Kutas 1978). The exponential decrease of the U, Th and K concentration for consolidated crust down to 35 km was taken into account (Rybach 1979). The value of 0.04 $\mu W/m^3$ (Kutas 1978) was accepted for the heat generation in the bottom layers of the earth's crust.

Results

Fig. 2 shows the profile under study. It summarizes both seismological and geothermal data. The earthquake foci for the period 1900–1975, located less than 25 km distant from the profile, are shown as projections on the figure plane. Vp velocities, fractures and reflectors are also depicted in the figure.

Tab. 1. The results of calculations of geotherms, mantle heat flow and thermal gradients of the cross-section Black Sea – Caspian Sea.

No of the unit	Moho temperature (in average) ($^{\circ}$C)	Mantle heat flow (in average) (mW/m^2)	Temperature at the base of sedimentary layer (maximum) ($^{\circ}$C)	Thermal gradient in crystalline crust (in average) ($^{\circ}$C/km)
I	650	17	120	13.2
II	670	17	100	12.8
III	550	12	200	10.5
IV	600	10	220	11.4
V	650	15	300	11.2
VI	700	19	280	13.7
VII	450	11	130	9.7

Some results of the calculations of geothermal fields are presented in Tab. 1. The Moho temperatures vary from 450 up to 720 $^{\circ}$C along the profile, with an average of 600 $^{\circ}$C. These results are similar to those of Buachidze (1979). The mantle heat flow varies from 8 to 19 mW/m^2, the mean value being 14 mW/m^2, that is, one-third of the surface heat flow along the cross-section. In some places the temperature at the base of a sedimentary layer reaches 300 $^{\circ}$C. For the greater part of the profile crossing the Kura basin this parameter equals 200 $\pm$ 50 $^{\circ}$C beneath the sedimentary layer 6–14 km thick.

The vertical geothermal gradient in crystalline rocks of the earth's crust is approximately equal to 12 ± 3 °C km for the regions under study.

Tab. 2. The distribution of the earthquake foci relative to geotherms.

Temperature range (°C)	Number of foci	In percentage
<100	2	2
100–200	29	30
200–300	31	32
300–400	27	28
>400	7	8

In Fig. 2 it can be seen that most of the foci are located above the depth of the 400 °C isotherm. Tab. 2, which illustrates the distribution of hypocentres in relation to isotherms, shows that over 90 % of the earthquakes with magnitude 2.5 and higher are concentrated in the earth's crust in the temperature range between 100 and 400 °C. As far as the events with smaller magnitude are concerned, they are known to tend to concentrate in the upper layers of the earth's crust (Nersesov et al. 1979), where the temperature is apparently lower than 400 °C for most regions of the Caucasus. It should be noted that about 85 % of all earthquakes have their hypocentres in the upper 10 km, and about 93 % in the upper 15 km of the earth's crust in the Garm region (Nersesov et al. 1979). Thus, the temperature of 400 °C can, to some extent, be viewed as critical for the occurrence of crustal earthquakes. Among the events whose foci occurred deeper in the earth's crust, the shocks with a magnitude of over 3.5 are predominant. The depth of the 400 °C isotherm is greater than the thickness of the "granite" layer along almost the whole profile. Fig. 2 indicates that less than 5 % of all the earthquakes in question have their foci in sediments. Thus, the reason for high seismicity in a cold medium (up to 400 °C) can be assumed to be fragility of "granite" layer rocks.

From the joint analysis of temperatures and Vp velocities of the cross-section, it follows that high-velocity inclusions correspond to higher temperatures at the Moho discontinuity beneath 630–720 °C, the depth of the latter being greater than in other parts of the profile and reaching 49 km.

Moreover, it should be noted that there is a correlation between the thickness of a sedimentary layer and the higher temperatures increasing with depth. In other words, the geotherms rise as the thickness of a sedimentary layer decreases. It is the case of the areas with almost the same values of surface heat flow. This effect appears to be the consequence of poorly conducting properties of a sedimentary layer. It is revealed by the inverse dependence of the coefficient of thermal conductivity on temperature.

Conclusions

1. The temperature of the volume containing earthquake foci is an essential parameter of a seismic region.
2. The calculated temperatures along the sublatitudinal (about 42°) Caucasus DSS cross-section vary in the range of 450–720 °C at the Moho

discontinuity, the corresponding mantle heat flow being 8-19 mW/m^2, respectively.

3. The majority of strong earthquakes (M>2.5) along the profile studied have their hypocentres in layers of the earth's crust at temperatures no higher than 400 °C. The account of the weaker earthquakes of the Caucasus and Garm regions seems to supplement the conclusion.

4. The zones of seismic Vp velocity anomalies are revealed thermal fields.

References

Artyushkov, E.V., 1972: The origin of high stresses in the earth's crust. - Izv. Acad. Sci. USSR. Physics of the Earth, 8, 1972: 3-25. (Russ.)

Buachidze, G.I. & Shaorshadze, M.P., 1974: Geothermal conditions of Western Georgia. - In: Geothermics, Moscow, Geol. Inst. Acad. Sci.: 37-39. (Russ.)

Čermák, V., 1979: Heat flow map of Europe. - In: V. Čermák & L. Rybach (Eds.), Terrestrial Heat flow in Europe. - Springer-Verlag, Berlin, Heidelberg, New York, 3-40.

Hurtig, E. & Stromeyer, D., 1979: Analysis of the Surface Heat Flow. Temperature Variations in the Upper Mantle Beneath Europe. - In: V. Čermák & L. Rybach (Eds.), Terrestrial Heat Flow in Europe. - Springer-Verlag, Berlin, Heidelberg, New York, 107-111.

Kondorskaya, N.V., Balavadze, B.K. et al., 1978: Some aspects of joint analyses of seismological data by the USSO stations and DSS for the Caucasus region. - Proceedings of the XVI Gen. Ass. of ESC., Strassburg.

Kunin, N.Ya. & Volvovsky, B.S., 1976: General concepts of the methods of geological interpretation of complex geophysical data. - In: Geophysical fields of the lithosphere of the earth. - Moscow, Sov. Radio, 16-19. (Russ.)

Kutas, R.I., 1978: Heat flow field and thermal model of the earth's crust. - Kiev, Naukova Dumka, 146 p. (Russ.)

— , 1979: A geothermal model of the earth's crust on the territory of the Ukrainian Shield.-In: V. Čermák & L. Rybach (Eds.), Terrestrial Heat Flow in Europe. - Springer-Verlag, Berlin, Heidelberg, New York, 309-315.

Milanovsky, E.E., 1968: Neotectonics of the Caucasus. - Moscow, Nedra, 484. (Russ.)

Nersesov, I.L., Ponomarjev, B.S. & Teytelbaum, Yu.M., 1979: The variations of the earth's crustal seismicity in different depth layers and seismic forecasting. - Dokl. Akad. Nauk SSSR, 247: 1100-1103. (Russ.)

New catalogue of strong earthquakes on the USSR territory, 1977. - Moscow, Nauka, 536 p.

Popov, V.V., 1963: On the relaxation of thermal stresses in the upper layers of the earth. - Izv. Acad. Sci. USSR, Geophys. 10: 31-36. (Russ.)

Rybach, L., 1978/79: The relationship between seismic velocity and radioactive heat production in crustal rocks: An exponential law. - Pure and Applied Geophysics, 117, 1/2: 75-82.

Zharkov, V.N., 1978: The internal constitution of the earth and planets. - Moscow, Nauka, 192. (Russ.)

Outlines for interpreting local heat flow anomalies in the Tuscan-Latial pre-Apenninic belt

Luciano Galeone and Francesco Mongelli

with 9 figures

Galeone, L. & Mongelli, F., 1982: Outlines for interpreting local heat flow anomalies in the Tuscan-Latial pre-Apenninic belt. - Geothermics and geothermal energy, eds. V. Čermák & R. Haenel, E. Schweizerbart'sche Verlagsbuchhandlung, Stuttgart: 181-190.

Abstract: The heat flow map of the Tuscan-Latial pre-Apenninic belt is strongly dominated by local anomalies. On the basis of the hydrogeological model by Calamai et al. (1977) we attribute these local anomalies to the tectonic features of the main aquifer (reservoir) and interpret them by simple heat conduction models.

Authors' addresses: L. Galeone, Istituto di Analisi Matematica, Università di Bari, 70100 Bari, Italy; F. Mongelli, Istituto di Geodesia e Geofisica, Università di Bari, 70100 Bari, Italy

Introduction

The Tuscan-Latial pre-Apenninic belt is the site of one of the highest surface heat flow values in the world. Intensive geothermal prospecting

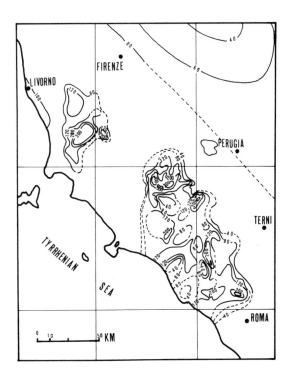

Fig. 1. Heat flow map of Tuscany and Northern Latium (in mW m^{-2}. According to Haenel et al. 1980).

The heat released by these bodies reaches by conduction and/or convection the aquifer where the fluid is moving from the recharge zone to an unknown discharge zone (Fig. 3). It is also very probable that thermal convection takes place in the fluid system or in sections thereof.

According to the fluid motions in the aquifer interior, and to the magma cooling, the temperature at the upper surface of the aquifer varies in space and time. The horizontal extent of the fluid system is presumably large in comparison with that size of the magma bodies; hence, the temperature in the impermeable cover of the confined aquifer is controlled only by the temperature of the upper surface of the aquifer itself. In this situation, the aquifer acts as an effective heat source, a "secoondary source" (Mongelli 1978): its regional trend controls the (anomalous) regional heat flow, and its local irregular structures produce local high heat flow.

Interpretation of local anomalies

It is important to note that negative anomalies are not interpretable by heat conduction models, and that local positive anomalies are considered to be those greater than 200 mW m^{-2}.

To test the model we can apply it to known cases: this means applying the heat conduction equation to situations of known geometry and verifying the consistency between observed and calculated temperatures or surface temperature gradients (or heat flow).

Fig. 2 gives some typical structures of the carbonate tectonics, such as sloping surfaces, faults, horsts, grabens, with which we associate the following geometries: wedge, step, double step.

It is generally assumed that the time required by magma bodies to reach the steady-state conditions is of the order of 2-300 000 years; furthermore this time is reduced by convective phenomena. Hence, a steady temperature regime on the upper surface of the aquifer may be assumed for the Tuscan-Latial pre-Apenninic belt. It is important also to note that positive anomalies are so high that eventual correction due to temperature perturbations of external origin may be considered as negligible by comparison.

Considering the problem in two dimensions, we may find solutions of the steady-state Fourier's equations by the Finite Differences Method; the domain where the equation has to be integrated is covered by a "mesh point" obtained by straight lines parallel to the x axis (the earth's surface) and the z axis (the depth) at distances Δx and Δz. For each internal point of the mesh (x_i, y_j), setting $T(x_i, y_j) = T_{i,j}$ we get

$$\frac{T_{i+1,j} + T_{i-1,j} - 2T_{i,j}}{\Delta x^2} + \frac{T_{i,j+1} + T_{i,j-1} - 2T_{i,j}}{\Delta x^2} = 0$$

Boundary conditions are the temperature at the boundaries of the structures. The problem is thus reduced to the solution of a system of linear equations which may be performed by the Relaxation Method.

Since this method gives the temperatures at each point, the temperature gradient or the heat flow near the surface may also be calculated and compared with the observed values.

First test: Sloping surface of the Travale field (wedge)

This is one of the best studied geothermal fields in Italy. With the use
of the results of geophysical and hydrogeological research and those of
deep temperature measurements in exploration and exploitation wells Fig. 4
was drawn (Calore et al. 1975). It represents the situation along a section
of the field. This case was studied by Galeone & Mongelli (1979); by using the
the data of Fig. 4 they obtained the temperature distribution on the
sloping surface (ξ in km)

$$f(\xi) = 250 - 235\ e^{-0.00121 \cdot \xi}\ (^{\circ}C)$$

moreover, they assumed the following boundary conditions for the wedge
ABC (Fig. 5)

on AB: $T_O = 15\ ^{\circ}C$

$\quad$ BC: $T = 15 + \dfrac{250-15}{BC}\ z\ (^{\circ}C)\quad 0<z<H$

$\quad$ AC: $T = f(\xi)\qquad\quad (^{\circ}C)$

Fig. 5 shows the numerical results obtained for the wedge; a comparison
with Fig. 4 shows the consistency between observed and calculated temperatures.

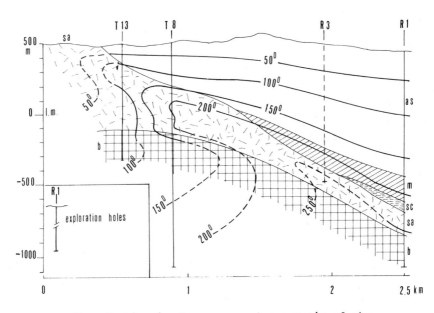

Fig. 4. Temperature distribution in the cover and reservoir of the
Travale field (in $^{\circ}C$). as = shaley clays; m = sandstones (macigno);
sc = shales; sa = anhydritic series; b = Palaeozoic basement (according
to Calore et al. 1979).

Second test: A fault of the Travale field (Step)

Fig. 6 shows another section of the field intersected by a fault, obtained
from observed data (Calore et al. 1979). Regardless of the slope of the
reservoir, we assume the model of Fig. 7 with the following boundary

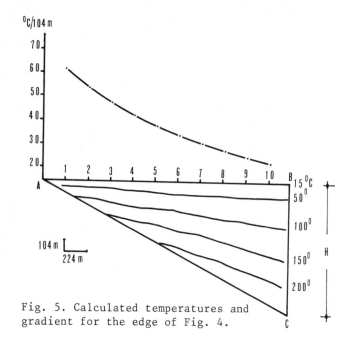

Fig. 5. Calculated temperatures and gradient for the edge of Fig. 4.

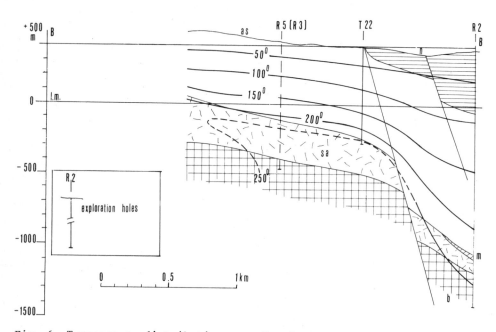

Fig. 6. Temperature distribution around a fault of the Travale field (in °C, according to Calore et al. 1979). Symbols are the same as in Fig. 4.

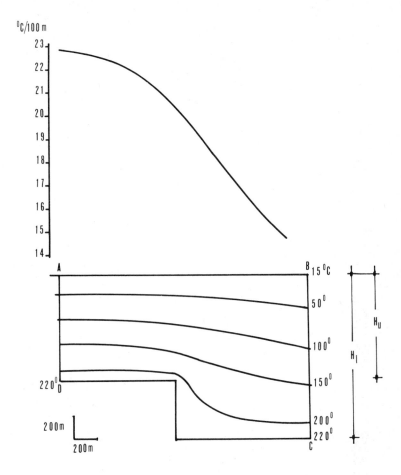

Fig. 7. Calculated temperatures and gradient for the step of Fig. 6.

conditions:

on AB: T_O = 15 ^{O}C

DC: T = 220 ^{O}C

AD: T = 15° + $\dfrac{220 - 15}{AD}$ z   (^{O}C) $0 < z < H_u$

BC: T = 15° + $\dfrac{220 - 15}{BC}$ z   (^{O}C) $0 < z < H_e$

Fig. 7 also shows the numerical results obtained for the polygon ABCD; the consistency with Fig. 7 is evident.

Third test: The horst of the Torre Alfina field (double step)

The horst is the most favourable structure for accumulating hot fluids.
    Fig. 8 shows a section of the Torre Alfina field and the observed surface heat flow based upon available information (Cataldi & Rendina 1973, Barelli et al. 1978, Cataldi pers. commun.). The model of Fig. 9 is

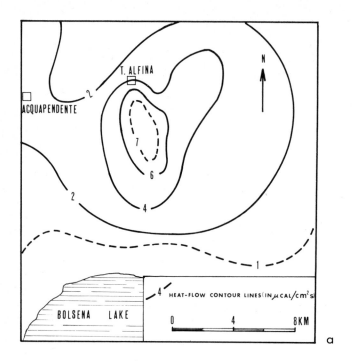

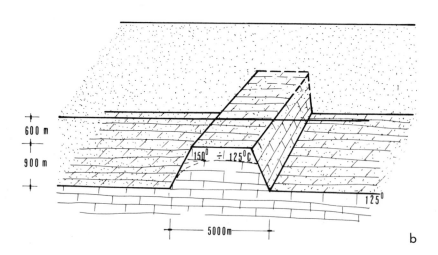

Fig. 8. a) Heat flow map of Torre Alfina field (according to Cataldi & Rendina, 1973).
b) Schematic section of the reservoir (according to Cataldi, pers. comm.).

is assumed to have the following boundary conditions:

on AH: T_o = 15 ^{o}C

   BC=FG: T = 125 ^{o}C

   DE   :   T = 150 ^{o}C

   AB=HG: $T = 15 + \dfrac{125 - 15}{BC} z$ (^{o}C), $0 < z < H_1$

   DC-FG: $T = 150 - \dfrac{150 - 125}{DC} z$ (^{o}C), $H_u < z < H_1$

Fig. 9 shows the numerical results obtained for the polygon ABCDEFGH, from which it is possible to deduce the consistency between observed and calculated heat flow.

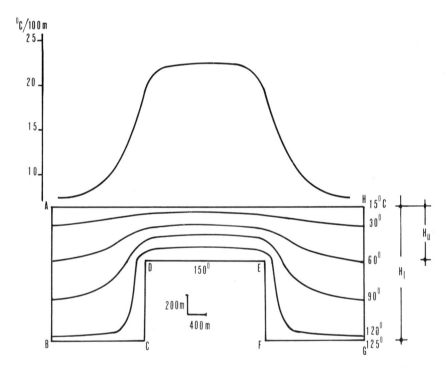

Fig. 9. Calculated temperatures and gradient for the double-step of Fig.8.

Conclusions

The consistency between observed and calculated local heat flow anomalies demonstrates that the temperature field in the cover is controlled only by the temperature distribution at the upper surface of the aquifer. This confirms the validity of the geothermal model of Fig. 3 in the sense that (I) all the heat flow coming from the interior is captured and transmitted by the fluid system (where the fluid is moving); (II) local positive anomalies are directly related to the tectonic structures. Thus, the knowledge of the geometry of the structures gained by other geophysical

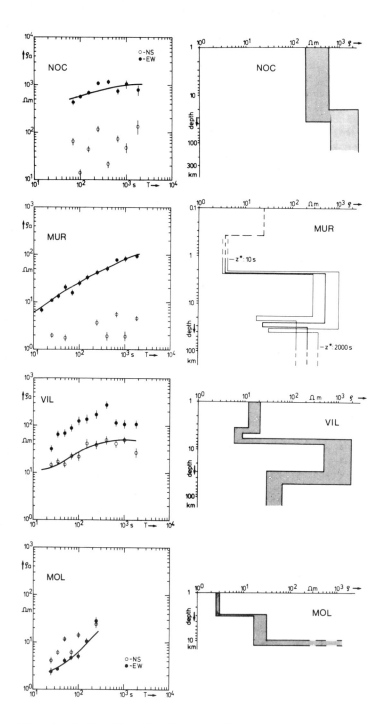

Fig. 2. *Left side:* The "measured" apparent resistivities are represented
as dots (full or open), the curve through these dots represents a
theoretically calculated apparent resistivity function, corresponding to
the resistivity-depth functions at the *right side* of the figure. The
error bounds of the resistivity models correspond to an assumed 10 %
scatter of the apparent resistivities.

resistivity and whether there may exist possibly more geothermal anomalies.

Results

The electric and magnetic fields which have been recorded at six sites have been selected at present for further processing and interpretation. The methods of analysis (frequency transformation and calculation of transfer functions) will not be discussed here. Generally two apparent resistivity curves resulted for a particular site due to the polarization of the electric field caused by lateral non-uniform resistivity distribution. These apparent resistivities are represented in Figs. 2 and 3 by the more or less scattering full and open dots. The coordinate systems have not yet been rotated into the main impedance coordinate system but are still NS and EW. However, one might readily infer from Figs. 2 and 3 that all apparent resistivities (except the one in NOC) point to rather low resistivities of the upper kilometers.

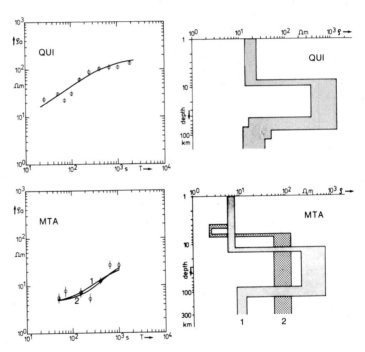

Fig. 3. Same as for Fig. 2.

A first tentative transformation of these apparent resistivities into resistivity-depth functions has been accomplished by an inversion algorithm of Schmucker (1974). These models are represented to the right in Figs. 2 and 3. It must be emphasized that these models are strictly only true if the subsurface is laterally uniform resistive. This is not the case at most of the sites. Nevertheless, the models calculated at NOC, MUR and QUI may represent an upper limit of the resistivities, the models for VIL and MOL may represent a lower limit. Clearly, at most of the sites at least 2-dimensional resistivity distribution could satisfy the two apparent resistivity curves.

Such 2-dimensional model calculations have not yet been started. But the main trend of the results shown here will not be affected crucially by a more sophisticated interpretation method.

Discussion and conclusion

The resistivity models of MUR, MOL and MTA which are close to the "hot areas" of Larderello, Travale and Monte Amiata display clearly the expected low resistivity values of the upper layers (see Fig. 4). Rather unexpected, however, are the low resistivities found also at sites distant from these known anomalies, as QUI and VIL. Also rather unexpected is the great thickness of the low resistive layer, except at NOC which might indicate the northern boundary of the geothermal area. Since the geologists predicted a (low resistive) sedimentary layer in this area of about 3 km thickness underlain by a crystalline basement, and the magnetotelluric results indicate a large amount of low resistive fluids partially down to 10 km one might conclude that the crystalline basement within Tuscany is fractured and filled with hot, circulating fluids. – There seems to exist a correspondence of thickness of the low resistive layer and the depth of reflection horizons (Batini et al. 1978). In a speculative model one might summarize both reflection and low resistivity results as effects of an allochthonous, crystalline nappe, giving rise to a primary heat source which feeds or heats the secondary, exploited heat sources in the known geothermal anomalies.

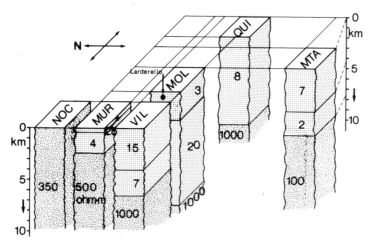

Fig. 4. A geographical block diagram of the resulting resistivity models. The main features are the strong increase of the thickness of the low resistive layer from N to S (NOC to VIL) and the remarkable thickness of this low resistive layer. Most probably this is due to deeply fractured allochthonous crystalline basement which allows the deep circulation of hot fluids.

Acknowledgements. We are greatly indebted to the magnetotelluric working group of the University of Munich (Dr. M. Beblo, Dr. A. Berktold, Dipl. Geophys. P. Wolfgram), who participated in the fieldwork. Also most of the magnetotelluric equipment was lend from this university. We also thank Prof. C. Morelli (Trieste), Prof. Norinelli (Padova) and Dr. S. Spitz (Padova) for their help. - This project has been supported by the Commission of the European Communities (contract no. 488-78-7EGD).

References

Batini, F., Burgassi, P.D., Cameli, G.M., Nicolich, R. & Squarci, P., 1978: Contribution to the study of deep lithospheric profiles: "deep" reflecting horizons in Larderello-Travale geothermal field. - Mem. Soc. Geol. Ital.

Schmucker, U., 1974: Erdmagnetische Tiefensondierung mit langperiodischen Variationen. - Prot. Koll. Erdmagn. Tiefensondierung. Grafrath: 313-343.

Relations between geothermal anomalies, deep groundwater flow and salinity distribution*

H. Scriba, M. Parini and D. Werner

with 4 figures

Scriba, H., Parini, M. & Werner, D., 1982: Relations between geo-
thermal anomalies, deep groundwater flow and salinity distribution.
- Geothermics and geothermal energy, eds. V. Čermák & R. Haenel,
E. Schweizerbart'sche Verlagsbuchhandlung, Stuttgart: 197-202.

Abstract: The mechanism of geothermal anomalies in the sedimentary
cover of the Rhinegraben is investigated, with special reference to
the area of Landau/Pfalz. The anomalies can be explained by uprising deep groundwater along faults in the basement. The source
depth of the water can be determined by numerical modeling. Moreover, the salinity distribution observed in the Landau field can
be explained.

Authors' address: Institut für Geophysik, ETH, CH-8093 Zürich,
Switzerland

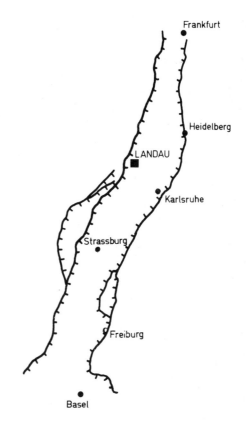

Fig. 1a. Location of Landau.

---

* Contribution No. 110 of the Institut für Geophysik, ETH Zürich.

Introduction

In this paper the mechanism of geothermal anomalies in the sedimentary
cover of the Rhinegraben is investigated. The upper Rhinegraben is geo-
thermally characterized by high subsurface temperatures (see e.g. Doebl
1970, Werner & Doebl 1974). A well known significant anomaly is located
in the oil-field of Landau/Pfalz (Fig. 1a). Our considerations here are
restricted to this case, but they can be applied to other anomalies as
well (e.g. Stockstadt, Soultz/Pechelbronn). Continuous temperature logs
have been carried out down to 1300 m. The results are described in Werner
& Parini (1980). In the centre of the anomaly the temperature at 1000 m
depth is about 100 °C: the corresponding heat flow is about 120 mW/m^2
(Fig. 1b).

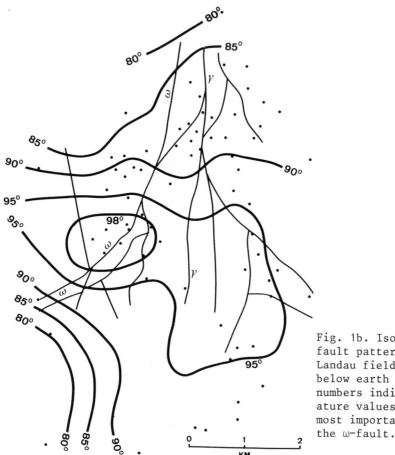

Fig. 1b. Isotherms and
fault pattern of the
Landau field at 1000 m
below earth surface. The
numbers indicate temper-
ature values in °C. The
most important fault is
the ω-fault.

Hydrothermal model

It is postulated that the observed thermal anomalies in the sediments of
the Rhinegraben are caused by water circulation from the crystalline basement. It has been shown that other attempts of interpretation do not agree
with the observations (Werner 1975).

A two-dimensional model, based on the assumption that some deep faults

in the basement are permeable, at least locally (Werner & Parini 1980),
has been computed. Groundwater rises along these faults, transmits its
heat to the surroundings and builds up a thermal anomaly.

The heat transport in a porous medium can be described by the differen-
tiel equation

$$\gamma \, \vec{v} \, \text{grad} \, T + \frac{\partial T}{\partial t} = D \, \Delta^2 \, T$$

where

T = temperature
t = time
$\vec{v}$ = vector of the Darcy filtration velocity
γ = ratio of the heat capacities of water and rock
D = "effective thermal diffusivity" in water-permeable zones, which
     includes the effect of thermal dispersion.

In our model the Darcy velocity field is a given input quantity and is
assumed to be constant with time. A finite difference method in space and
time was used for computation.

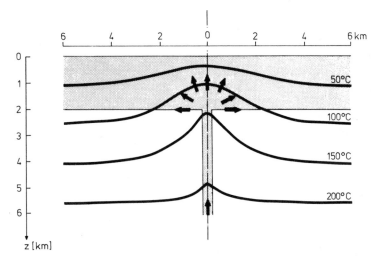

Fig. 2. Two-dimensional hydrothermal model. Deep groundwater rises along
a fault zone and builds up a thermal anomaly. The shaded zones are
assumed to be permeable; the arrows indicate the direction of flow. The
whole sedimentary cover is assumed to be permeable with a radial flow
distribution. The total value of the deep groundwater flow is $S = 50 \text{ m}^2/\text{a}$
(= 50 m³/a for 1 m of length perpendicular to the figure plane).

Fig. 2 shows a result of the model computations, which is in good agreement with the temperature distribution in the Landau field. The total
groundwater flow amounts to 50 m³/a.m, which is a very small value. It can
easily be provided by surface water from a drainage area, possibly outside
the graben margins (e.g. Pfälzerwald). The observed anomaly can be explained by this model with a source depth $z_0 = 6000$ m. If $z_0 < 6000$ m is
used the calculations cannot fit the anomaly if the same order of magnitude
of the flux is supposed. Calculations show that the temperature distribu-

tion becomes nearly stationary after about 100 000 years.

Salinity

The rise of groundwater from the basement should affect the salinity
distribution. One can assume that the rising water has a relatively low
salinity compared with the formation water in the graben fill. The sediments are characterized by an increasing salinity of the formation water
with depth (see e.g. Heling 1969): the mean gradient amounts to about
10 g/1·(100 m). In the Landau field, however, there is a marked freshwater
zone near the prominent ω-fault (Schad 1962). From NW to SE the salinity
increases from 20 g/1 to more than 100 g/1 within a few kilometers. It is
remarkable that the freshwater zone as well as the centre of the thermal
anomaly are connected with the ω-fault.
    These observations can be further confirmed by looking at the electrical
resistivity distribution. Fig. 3 shows a resistivity section across the
ω-fault. A high resistivity anomaly can be clearly recognized near the
fault. According to Wyllie (1963), the Landau temperature anomaly causes
a resistivity change by a factor of 1.5. The salinity variation by a
factor of 5 would also cause a resistivity change by a factor of 5. Thus,
salinity plays the dominant role. Moreover, the increased temperature
would lead to a resistivity low.

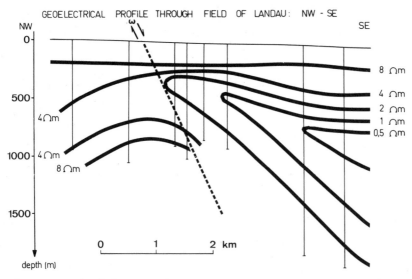

Fig. 3. Resistivity section across the ω-fault. The vertical lines indicate boreholes.

A salinity model, in analogy to the hydrothermal model, has been computed
in order to simulate the observed salinity anomaly (Parini et al. 1980).
Fig. 4 shows a result of a two-dimensional calculation with the identical
velocity distribution and the same time span (100 000 a) as in the model
of Fig. 2. The original salinity increases linearly with depth (0-100
units). The low-salinity zone around the fault is evident: it has a diameter of about 2 km, which agrees with the observations of Schad (1962)
and with the resistivity anomaly (Fig. 3).

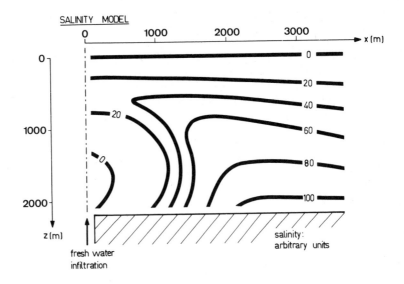

Fig. 4. Two-dimensional, symmetrical salinity model corresponding to the water flow distribution of the hydrothermal model in Fig. 2. The undisturbed salinity distribution (the palaeosalinity) increases linearly with depth (concentration in arbitrary units).

Of course, this model and the hydrothermal model as well represent only a rough approximation to the conditions in the sediments. In reality, impermeable zones exist, at least in connection with the oil traps: this should be considered by more sophisticated models.

Acknowledgements: This study was carried out under contract with the Commission of the European Communities (Project no. 321/79/2 EGD). We thank Wintershall A.G. for granting permission to carry out the measurements and especially Dr. F. Doebl for providing all necessary information.

References

Doebl, F., 1970: Die geothermischen Verhältnisse des Oelfeldes Landau/ Pfalz. - In: J.H. Illies & St. Müller (Eds.), Graben Problems. - Schweizerbart, Stuttgart, 110-116.

Heling, D., 1969: Relationships between initial porosity of Tertiary argillaceous sediments and palaeosalinity in the Rhinegraben (SW-Germany). - J. Sediment. Petrol. 39: 246-254.

Parini, M., Scriba, H., Sieber, C. & Werner, D., 1980: Geothermal anomalies in the Rhinegraben sediments and their explanation by uprising deep groundwater from the crystalline basement. - Proceedings of the Seminar on Geothermal Energy, Strasbourg, 4.-6.3.1980 (in press).

Schad, A., 1962: Das Erdoelfeld Landau. - Abh. Geol. Landesamt Baden-Württemberg, 4: 81-101.

Werner, D., 1975: Probleme der Geothermik am Beispiel des Rheingrabens. - Thesis, University of Karlsruhe.

Werner, D. & Doebl, F., 1974: Eine geothermische Karte des Rheingraben-
untergrundes. - In: J.H. Illies and K. Fuchs (Eds.), Approaches to
taphrogenesis. - Schweizerbart, Stuttgart, 182-191.
Werner, D. & Parini, M., 1980: The geothermal anomaly of Landau/Pfalz:
an attempt of interpretation. - J. Geophys. 48: 28-33.
Wyllie, M.R.J., 1963: The fundamentals of well log interpretation. -
Academic Press, New York, 238 pp.

Main types of thermal water reservoirs in the Pannonian Basin

K. Korim

with 5 figures

Korim, K., 1982: Main types of thermal water reservoirs in the
Pannonian Basin. - Geothermics and geothermal energy, eds. V.
Čermák & R. Haenel, E. Schweizerbart'sche Verlagsbuchhandlung,
Stuttgart: 203-210.

Abstract: Although there is a variety of thermal water reservoir
systems, only two types are of regional character in the Pannonian
Basin. One is the Upper Pannonian (Middle Pliocene) deep-lying,
confined multilayer or multiunit reservoir system consisting of
horizontal or quasi-horizontal sand beds; the other is the
fractured, fissured, jointed and partly karstified carbonate rock
complex of Triassic age. A wide range of geological factors, such
as subcrustal magmatic processes, tectonics, basin evolution and
sedimentation are responsible for the occurrence, patterns,
characteristics and properties of the aquifers and geothermal
resources. The majority of the thermal water reservoirs are sealed
and excluded from the hydrological cycle and have no recharge. On
the contrary, shallow aquifers are more or less connected to the
hydrological cycle and are recharged.

Author's address: Water Prospecting and Drilling Company, Budapest,
Hungary

Introduction

There are different kinds of thermal water reservoir systems in the
Pannonian Basin within a regional positive geothermal anomalous environment. Various geological-hydrogeological conditions govern the size,
extension, dimension, water-yielding capacity and general hydrodynamical-
hydrochemical features of the aquifers and of the thermal water resources.
    It is well known that the geothermal conditions of the Carpathian Basin
are determined primarily by a particular crustal structure and by magmatic-
kinetic processes taking place within the earth's crust and the upper
mantle (Stegena 1973). Secondarily, however, it is the geological setting
that controls the geothermal phenomena.
    The Hungarian Basin consists of three major geological units:
- The Precambrian and Palaeozoic basement with rigid, highly deformed,
  metamorphic crystalline rocks; it represents the framework and infra-
  structure of the whole Carpathian Basin. Only its carbonate members
  have some locally developed fractured-fissured aquifer systems (e.g.
  the Devonian dolomite thermal water reservoir at Bük in West-Hungary).
- A superimposed Mesozoic basement composed mainly of thick and extended
  limestone and dolomite formations including considerable karstic and
  fissure water resources.
- Neogene (mainly Pliocene) clastic rock sequences including several
  hydrostratigraphic horizons with formation water. This vast sedimentary
  formation represents the superstructure of the Carpathian intermontane
  region (Korim 1972).

Main types of thermal water reservoirs

1. Deep-lying confined multiple sandy reservoirs of horizontally stratified character

In the great Hungarian sedimentary basin, which consists of several sub-basins, superimposed composite sand and sandstone bodies occur alternately with clay, clayey marl, marl and silt beds. That is, intricate series of aquifer, semiaquifer, aquitard and aquiclude formations were developed as a result of lacustrine and delta-like sedimentation in the Pliocene and the Quaternary. The spatial distribution and external geometry of sand and sandstone units are defined by the general character and pattern of the repetitive sedimentary sequence.

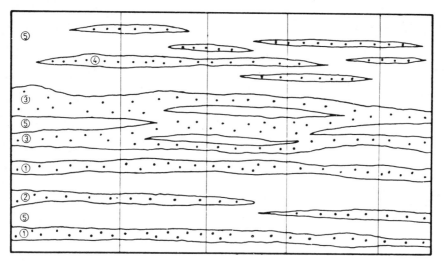

Fig. 1. Basic forms of sand body development.
1. Continuous layer, 2. Wedged or pinched-out layer, 3. Coalescing layers, 4. Sand lenses, 5. Impervious layer.

The Upper Pannonian, e.e., the Middle Pliocene sedimentary column includes the most important thermal water reservoir system in Hungary as well as in the conterminous basin areas of the neighbouring countries (Czechoslovakia, Yougoslavia). It is a multilayer or multiunit or multiple reservoir system. It ranges in depth from 600 metres to 2500 metres and contains static-stagnant formation water which is overwhelmingly sealed and excluded from the hydrological cycle. Within these horizontal or sub-horizontal sand and sandstone deposits the reservoir pressure is everywhere equal or near to the hydrostatic value. The base or reservoir temperature varies from 45 to 140 °C. The highest outflowing temperature of the thermal water is about 100 °C in the region surrounding Szentes and Orosháza (SE-Hungary). The total content of dissolved solids in the thermal water ranges from 1500 to 6000 ppm and with a few exceptions is predominantly of alkaline hydrogencarbonate character. It is remarkable that some gas is dissolved in the thermal water nearly everywhere this is one of the most important reservoir energy sources.
    The Upper Pannonian multiple reservoir system contains the greatest

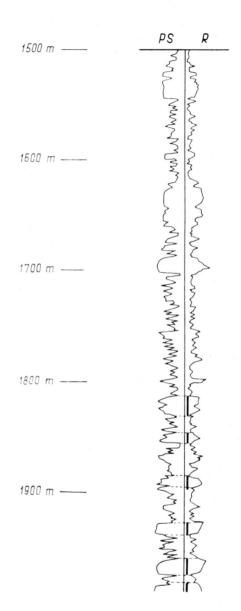

Fig. 2. General pattern of a typical
Upper Pannonian multiunit-multistorey
thermal water reservoir system with
multiple production sections
(according to an electrical log).

thermal water resources and reserves in Hungary. About 70 per cent of all
the thermal wells produce from these formations. The thermal wells are
completed as multiple production wells. Their water-yielding capacity
ranges from a few hundred to 2000 - 3000 litres per minute at 0.1 to 0.2
bar well-head pressure. The initial static or shut-down well-head pressure
is about 4 to 5 bar. The relatively high water temperature allows a wide
range of utilization for balneological - therapeutical, agricultural and
space-heating purposes. A striking feature of this thermal water is the
multipurpose utilization, especially in SE-Hungary and in W-Hungary
(Boldizsór 1975, Boldizsór & Korim 1975).

2. Semi-confined stratified sandy reservoirs

In some depressed areas of the Great Hungarian Plain Levantinian sand
layers and Quaternary fluviatile sand and gravel deposits within a depth
interval of 400 to 1300 m also include thermal water resources at lower
temperatures (ranging from 25 to 55 °C, as measured at the well-head).
These thermal water-bearing formations also represent multi-unit or
multiple reservoir systems. The shallower members of this group communicate
with the surface and with the hydrological cycle and consequently a certain
recharge takes place.

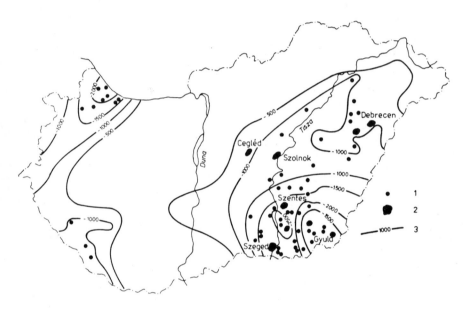

Fig. 3. Thermal water wells with more than 60 °C outflowing temperature
producing from the Upper Pannonian reservoir system.
1. Thermal water well, 2. Group of thermal water wells, 3. Contour line
of the bottom part of the Upper Pannonian sequence (below sea level in
metres).

3. Composite leaky thermal water reservoirs

A well-known and closely investigated example of this type of reservoir is
situated at Tiszakécske along the Tisza River. A water circulation system
is operating here, i.e., a water exchange of convective character occurs
through a stratigraphic window within the Pliocene and Quaternary sedimentary sequence at a depth interval of 50 to 1000 metres. The ascending
branch of this system creates a narrow belt of a remarkable positive
geothermal zone while south of this area the descending branch is
characterized by relatively low temperatures (Alföldi et al. 1978).

4. Overpressured thermal water reservoirs

Within the deepest part of the Neogene sedimentary basins Miocene clastic
sediments include some thermal water reservoirs of limited dimensions and

extension. These aquifers contain fossil saline formation water and are characterized by high reservoir pressure (so-called overpressured reservoirs). The reservoir temperature within these deep-lying Miocene basin sediments ranges from 160 to 250 °C. These reservoirs are totally excluded from the hydrological cycle, and because of their limited extent, the depletion is very quick and the life-time of such thermal water reservoirs is quite short.

5. Deeply buried fractured-fissured carbonate reservoir system

The Mesozoic carbonate basement rock complex is a very important thermal-water-bearing formation in Hungary. The buried portion of these carbonate masses is overlain by a Tertiary sedimentary sequence of differing thickness, and has a great regional extension. The upper part of the carbonate complex is intensively karstified not only on its outcropping parts but also in its buried portions (palaeokarsts). The rigid, brittle limestones, dolomites and marlstones, mainly of the Triassic age, include several areally separated thermal water reservoir units as a result of their vertical or subvertical fractures, fissures, joints of varying sizes and frequencies.

The deep-seated buried carbonate masses with well-developed fracture patterns and of secondary porosity enclose such thermal water reservoirs which have no cummunication with meteoric water and have no recharge. The fissure and karstic thermal water has a high concentration of dissolved solids and relatively high temperature (about 80 to 100 °C measured at the well-head). The most significant occurrences of this type of thermal water reservoirs is at Zalakaros, Igal and Táska in Transdanuvia.

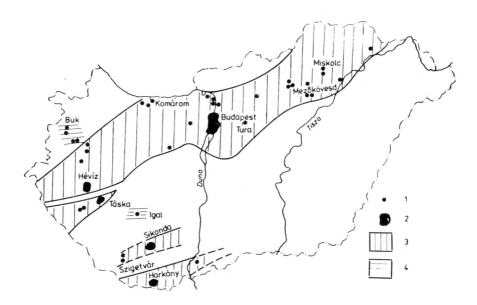

Fig. 4. Thermal water wells completed in carbonate rocks with more than 35 °C outflowing temperature.
1. Thermal well, 2. Group of thermal wells, 3. Triassic rocks,
4. Palaeozoic rocks.

6. Shallow fractured-jointed-karstified carbonate reservoirs

The shallow, near-surface carbonate thermal water reservoirs have a wide
variety because of the highly anisotropic character of the fracture-
fissure pattern. They are sometimes associated with major fault lines,
and along such zones very important natural thermal springs issue at the
surface, as for example, along the Buda Thermal Line. The extent, dimension
and configuration of the fractured reservoirs, along with the reservoir
properties differ widely in the function of the burial, the position and
mode of occurrence of the flow-path as well as the recharge conditions.
In the near-surface environment the hydrological cycle exerts a decisive
influence upon the water budget as well as upon the water chemistry of the
fractured-fissured aquifers. In the deeper reservoirs, however, flow,
exchange and mixing with meteoric water generally take place very slowly.
As a result, near-surface fissure water or karstic water is usually of
low concentration, representing the Ca-Mg-hydrogencarbonate type.
   The real temperature conditions of the carbonate reservoirs can be only
inferred from the flowing water temperature measured at the well-head,
since the base temperature cannot be directly measured because of the
vertical character of the flow-path system within the fractured-fissured
rock body. As a result of the varying, intricate and highly anisotropic
pattern of the carbonate thermal water reservoir system, the outflowing
water temperature may differ greatly within a limited area, too (e.g. in
Budapest it ranges from 23 to 75 $^{\circ}$C). The ascending and descending limbs
of some convective flow systems within the vast carbonate complex greatly
affect the temperature conditions (Alföndi et al. 1968).
   The water-yielding and storage capacity of carbonate reservoirs depends
primarily on the development of the fracture system and consequently, the
yield of the thermal water wells completed in carbonate rocks varies
within wide ranges and may occasionally be very high (e.g. 25 000 litres/
minute in Budapest).
   Thermal water from the carbonate reservoirs is utilized mainly for
balneological-therapeutical purposes and a highly developed balneological
network has been established in the whole country.

7. Composite bedrock aquifers

In some places different types and geologically distinct units of superimposed brittle, hard, fractured-fissured-jointed rocks form a common
thermal water reservoir of a uniform hydrodynamic system because of the
interconnecting fracture network. This type of composite reservoir occurs,
for example, in Budapest and in Héviz, where the real reservoir is the
underlying Triassic carbonate complex, whereas the overlying Eocene or
Pliocene fractured formation is only the outlet of this combined system.
   It must be noted that apart from the above mentioned main types of
thermal water reservoirs there are also other reservoirs which are,
however, of minor importance and extent (Bélteky et al. 1966, 1971, 1977).
   All of these different types of thermal water reservoirs clearly
exemplify the important and decisive role of different geological settings
and environments affecting the development and formation of the thermal-
water-bearing units and horizons and the associated geothermal processes
in the Pannonian Basin.

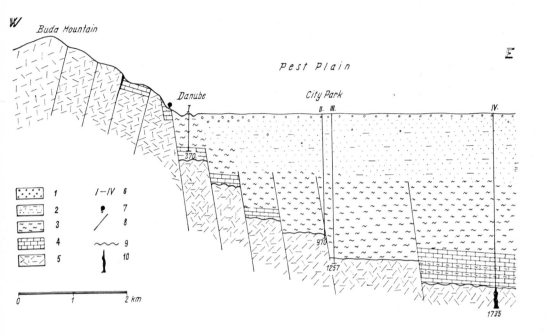

Fig. 5. Cross section over Budapest thermal line and thermal water
reservoirs.
1. Alluvial gravel, 2. Pliocene and Miocene sand, clay and clayey marl,
3. Oligocene marl, 4. Eocene limestone, 5. Triassic carbonate rocks,
6. Thermal water well, 7. Thermal spring, 8. Fault, 9. Unconformity,
10. Palaeokarstic cavern.

References

Alföldi, L. et al., 1968: Thermal Wells of Budapest. - Publ. of Research
    Inst. Water Resources Development, Budapest.
Alföldi, L., Erdélyi, M., Gálfi, J., Korim, K. & Liebe, P., 1978: A geo-
    thermal flow system in the Pannonian Basin, case history of complex
    hydrogeological study at Tiszakécske. - Proceedings of the Symposium
    on Hydrogeology of Great Sedimentary Basins, IAHS. Ann. Inst. Geol.
    Publ. Hung., 59/1-4, pp. 716-732.
Bélteky, L. et al., 1966, 1971, 1977: Thermal Wells of Hungary. - Publ. of
    Research Inst. Water Resources Development, Vols. 1, 2, 3, Budapest.
Boldizsár, T., 1975: Research and development of geothermal energy pro-
    duction in Hungary. - Geothermics, 4: 44-56.
Boldizsár, T. & Korim, K., 1975: Hydrogeology of the Pannonian geothermal
    basin. - Proceedings, 2nd U.N. Symposium on the Development and Use of
    Geothermal Resources. - San Francisco, pp. 297-303.
Korim, K., 1972: Geological aspects of thermal water occurrences in
    Hungary. - Geothermics, 1: 96-102.
Stegena, L., 1973: Cenozoic evolution of the Pannonian Basin. - M.T.A.
    X. Oszt. Közl., 6/1-4: 257-265. (in Hungarian)

Oxygen and hydrogen isotopic composition of water in the Puga and Manikaran geothermal areas (Himalaya), India

B. Kumar, K. S. N. Sahay, A. K. Baksi and M. L. Gupta

with 4 figures and 2 tables

Kumar, B., Sahay, K.S.N., Baksi, A.K. & Gupta, M.L., 1982: Oxygen and hydrogen isotopic composition of water in the Puga and Manikaran geothermal areas (Himalaya), India. - Geothermics and geothermal energy, eds. V. Čermák & R. Haenel, E. Schweizerbart'sche Verlagsbuchhandlung, Stuttgart: 211-218.

Abstract: The stable isotopic studies of water of Puga and Manikaran geothermal areas indicate that the water is predominantly of meteoric origin. δD contents of Puga thermal water are more or less similar to that of local meteoric water; however, $\delta^{18}O$ values show an oxygen shift of about 3 ‰o towards positive values, which is generally a characteristic feature observed in deep water circulating through high temperature environments. A potential zone for further investigations has been suggested on the basis of observed maximum oxygen shift. Isotopic ratios of Puga water and the topography of the area suggest that area of recharge for Puga thermal fluids may lie west of Puga. The thermal water in Manikaran is a mixture of a local cold meteoric component and a hot thermal water component, also of meteoric origin but from higher elevations.

Authors' address: National Geophysical Research Institute, Hydera Hyderabad, 500 007, India

Introduction

Puga and Manikaran form parts of Himalayan Geothermal Province where profuse hot spring activities occur. These areas have been recognised as most promising for development of geothermal energy resources in India (Krishnaswamy 1976, Gupta et al. 1976a). Various geoinvestigations have been carried out in these areas over the last five to six years. Gupta (1974), Gupta & Sukhija (1974) and Ravi Shankar et al. (1976) reported the association of the spring water of Puga with magmatic components. The studies of Gupta et al. (1976b, 1979) showed that the hot spring water of Manikaran is a mixture of cold meteoric and deep thermal water.

The problem of the origin of geothermal water presently is best answered by stable isotopic studies. Craig (1963) applied these techniques to many geothermal areas of the world and found that meteoric water overwhelmingly dominates recharge of most geothermal systems.

In the following it is reported on results of preliminary isotopic study carried out on geothermal waters of Puga and Manikaran areas.

Location, geology and climate

Hot springs at Puga manifest in a narrow valley located at an altitude of 4000-4400 m above mean sea level in the Ladakh district of Jammu and

Kashmir state in northwest Himalaya. The Valley with a maximum width of about 1 km, trending eastwest, extends about 15 km between Sumdo on the east and Polokongla on the west. It lies towards south of the well known Indus Suture Zone. The valley floor is covered with alluvium, sandy soil, salt encrustations, glacial moraine and spring deposits. The main rock sequence in the area belongs to the Puga formation which exposes para-gneisses, carbonaceous phyllites interlayered with bands of limestone, schist and garnet-mica schist with amphibolite sills and dikes. To the east of the Puga formation the Sumdo formation is exposed, while in the west it is intruded by Polokongla granite.

The Manikaran (Lat. 32°N, Long. 77°21'E) hot spring area is situated in the Parbati valley at an altitude of 1600-1700 m. Besides Manikaran, hot springs are manifested at four other locations, viz. Khirganga, Pulga, Kasol and Jan in this valley. The main rock formation is a thick sequence of quartzite with minor phyllite and slates, which is underlain by grey and green phyllite with bands of slates and carbonaceous schists. Over-lying the quartzite are gneisses and schists of the Kulu formation (Chaturvedi & Raymahashay 1976).

The mean annual temperature at Puga is around 0 $^{\circ}$C. The Puga area receives very little rainfall but a substantial amount of snowfall. The Manikaran area receives a substantial amount of snowfall as well as rain-fall. The mean rainfall at Manikaran amounts to 1000-1200 mm.

Sampling details and experimental techniques

At Puga the sampling was done during September '74 and June '78 and at Manikaran during March-April, 1977. Sample locations are given in Figs. 1 and 2.

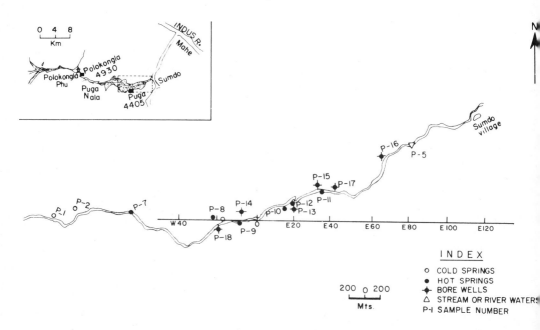

Fig. 1. Sample location map of the Puga geothermal area.

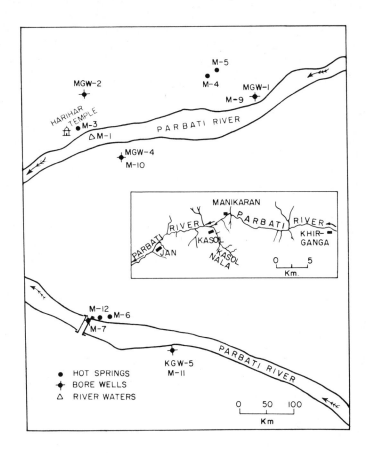

Fig. 2. Sample location map of the Manikaran-Kasol geothermal area.

Isotopic analyses of water samples were carried out by using established techniques. The oxygen isotopic composition of water was determined on CO_2 gas obtained by use of CO_2 equilibration method and hydrogen isotopic composition on hydrogen gas produced by reduction of water over uranium at 750 $^\circ$C. Further details are given elsewhere (Kumar et al. 1980). The isotopic data are expressed as deviation (δ) in parts per mil of the ^{18}O : ^{16}O and D : H ratios in the sample with respect to V-SMOW (Vienna – Standard Mean Ocean Water). The measuring precision for $\delta^{18}O$ is ± 0.1 ‰ and for δD it is ± 1 ‰.

Results

Puga geothermal area

The isotopic composition of water together with water temperature, date of collection and a short description of sampling points are given in Tab. 1. Isotopic data are plotted in Fig. 3. The data of Tab. 1 (excluding sample no. P-6) demonstrates that isotopic ratios of Puga cold and thermal water vary from -13.4 ‰ to -16.6 ‰ in case of oxygen and -115 ‰ to -123 ‰ for hydrogen. Sample No. P-6 is from Indus river near Mahe bridge and was

Tab. 1. Isotopic composition of water of the Puga geothermal area.

Sample number	Date of collection	Temp. °C	Short description	δD ‰	δ¹⁸O ‰
P-1	June, 1978	-	Cold spring-I	-116.2	-16.57
P-2	June, 1978	-	Cold spring-II	-117.5	-16.47
P-3	June, 1978	-	Polokongla cold water-I (2 km west of P-1)	-117.0	-16.30
P-4	June, 1978	-	Polokongla cold water -II (15 km extreme west of P-1)	-117.8	-
P-5	Sept., 1974	-	Puga stream water	-115.3	-15.33
P-6	June, 1978	-	Indus river near Mahe bridge (16 km away from Sumdo confluence)	-105.9	-14.29
P-7	June, 1978	47	Hot spring No. 33	-121.9	-15.82
P-8	June, 1978	52	Hot spring No. 45	-121.9	-15.30
P-9	June, 1978	57	Hot spring No. 40	-120.7	-16.35
P-10	June, 1978	70	Hot spring No. 24	-122.1	-15.40
P-11	June, 1978	58	Hot spring No. 27	-120.3	-15.30
P-12	Sept., 1974	-	Hot spring near borewell GW-2	-120.7	-13.96
P-13	Sept., 1974	125	Borewell GW-2	-118.7	-13.40
P-14	Sept., 1974	100	Borewell GW-5	-122.7	-14.80
P-15	Sept., 1974	-	Borewell GW-7	-116.6	-13.98
P-16	June, 1978	85	Borewell GW-8	-123.5	-15.20
P-17	June, 1978	100	Borewell GW-11	-121.3	-14.77
P-18	June, 1978	-	Borewell GW-22	-120.6	-14.77
P-16(a)	Sept., 1974	85	Borewell GW-8	-120.4	-15.14
P-17(a)	Sept., 1974	100	Borewell GW-11	-119.2	-14.94

collected about 16 km away from Sumdo confluence. The isotopic ratios of sample numbers P-16(a) and P-17(a) are not plotted in Fig. 3, but are included in the Tab. 1 in order to show that the ratios of thermal water from September, 1974 to June, 1978 have not changed significantly.

Manikaran geothermal area

Isotopic data of thermal and cold water together with their discharge temperature, chloride content, and a short description of samples are given in Tab. 2. Fig. 4 shows a plot of δD vs δ¹⁸O contents of Manikaran geothermal water. δD contents of Manikaran water varies from -69 ‰ to -49 ‰ and δ¹⁸O from -10.2 ‰ to -6.9 ‰. Parbati River water has the

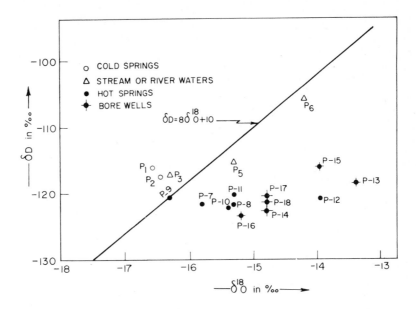

Fig. 3. Isotopic composition of water samples from the Puga geothermal area.

Tab. 2. Hydrogen and oxygen isotopic data of water of the Manikaran-Kasol geothermal area. Date of collection: March-April, 1977.

Sample number	Temp. $^\circ$C	Cl ppm	Short description	δD %o	$\delta^{18}O$ %o
M-1	–	31	Parbati river near Harihar temple	-64.5	-9.93
M-2	–	16	Parbati river near Jan hot spring	-68.7	-10.21
M-3	93	156	Manikaran hot spring No. 7	-56.6	-8.85
M-4	77.5	110	Manikaran hot spring No. 4	-58.6	-8.83
M-5	88.5	156	Manikaran hot spring No. 12	-58.9	-8.75
M-6	75.5	62.5	Kasol hot spring No. 3	-58.9	-8.72
M-7	74.5	62.5	Kasol hot spring No. 3B	-58.9	-9.05
M-8	33.5	360	Jan hot spring	-58.8	-8.62
M-9	80	62	Manikaran borewell MGW-1	-60.1	-9.18
M-10	45	78	Manikaran borewell MGW-4	-49.4	-7.13
M-11	52	62	Kasol borewell KGW-5	-57.7	-8.39
M-12	–	78.5	Kasol borewell	-48.6	-6.88

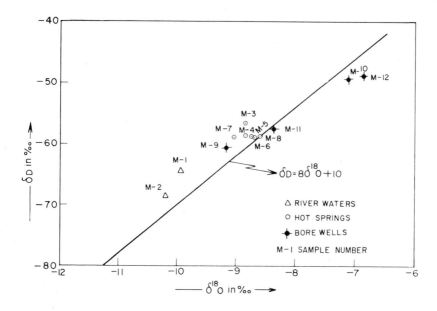

Fig. 4. Hydrogen and oxygen isotopic data for waters from the Manikaran-Kasol geothermal area.

most depleted isotopic values and water from one borehole in Manikaran and Kasol each has the most enriched values amongst all the samples.

Interpretation

The interpretation of the results is based on following considerations:
1. The delta values for deuterium and oxygen-18 for meteoric water all over the world are linearly correlated:

$$\delta D = 8 \, \delta^{18}O + 10.$$

This line is called the meteoric water line (Craig 1961).
2. The isotopic ratios of precipitation samples are determined by the temperature at the time of condensation, which in turn depends upon the altitude and latitude. At a given latitude, precipitation at higher altitudes is depleted in heavy isotopic species, as compared with precipitation occurring at lower altitudes (Dansgarrd 1964, Friedman et al. 1964).
3. The water occurring as steam or hot springs in most geothermal areas of the world has about the same deuterium concentration as the local meteoric water, but shows a characteristic shift of oxygen towards an $^{18}O : ^{16}O$ ratio higher than that of local meteoric water.

On the basis of the above considerations, isotopic data for Puga and Manikaran regions show that all the cold and hot water, described in Tab. 1 and 2 predominantly is of meteoric origin.

δD and $\delta^{18}O$ contents of Puga cold spring water are similar to that of Polokongla cold water I and II (Tab. 1). The deuterium content for Polokongla cold water II is -118 ‰ and the mean value for the deuterium content for Puga thermal water is -121 ‰. The difference observed in the deuterium contents of these samples reflects the "altitude effect" as mentioned above. The Polokongla cold water II was collected at an altitude of about 4750 m. The hot springs and bore water samples have depleted δD values as compared to those for Polokongla cold water, except in the case of sample no. P-15, whose isotopic values have been altered, most likely by evaporation. The variations in δD content of precipitation with altitude in the Himalayas are not well known but could be taken as 2 to 3 ‰ per 100 m. On this basis, it appears that the thermal water may be recharged mainly from precipitation falling at altitudes over 4800 m. Combining this inference with the topography of the area, it is most likely that the area of recharge for thermal water may be located west of Puga.

The oxygen shift in the case of Puga thermal water shows that the water is circulating at great depths and is subjected to high temperatures. The GW-2 borehole and a spring near the GW-2 borehole (sample no. P-12) show a maximum shift of oxygen by about 3 ‰. GW-2 also has the highest "base" temperature and chloride content and hence, its water may be most representative of the deep geothermal fluid.

Unlike Puga, in case of the Manikaran, the variations in $\delta^{18}O$ contents of thermal water are followed by variations in δD contents and no visible oxygen shift is observed. Friedman et al. (1964) gave a relationship between δD contents of natural water as a function of latitude and altitude. By combining the results of their relationship with isotopic results and chloride contents of Manikaran thermal and cold water, we infer:

1. that the thermal water of hot springs and boreholes is a mixture of a deep thermal water component and a local cold meteoric component from the Parbati River and/or from precipitation falling at an altitude of 1600-1800 m,
2. that the deep thermal water is recharged at altitudes higher than that of the discharge areas of hot springs and boreholes,
3. that the local cold meteoric component responsible for dilution of bore water of samples number M-10 and M-12 (Tab. 2) has originated mainly from rainfall occurring around the Manikaran and Kasol areas.

Conclusions

1. Meteoric water is a dominant component of thermal water of the Puga and Manikaran areas.
2. The thermal water of Puga probably recharged west of Puga and shows an oxygen isotopic shift, which is a characteristic feature of deep circulating high temperature fluids. The maximum oxygen shift has been observed in the water of the GW-2 borehole and of a spring near it thus indicating a promising zone for future investigations.
3. Deep thermal water of the Parbati valley is diluted with Parbati River water and/or water from local precipitation in various proportions before being discharged through hot springs and boreholes.

Acknowledgements. We gratefully acknowledge the help of Messrs. S.B. Singh, S.R. Sharma and Yadgiri for the field collection of samples and the assistance of Mr. G.G. Kingi in performing oxygen and deuterium analyses. Our special thanks are due to Dr. S. Balakrishna, Director, National Geophysical Research Institute, for granting permission to publish this paper.

References

Chaturvedi, L.N. & Raymahashay, B.C., 1976: Geological setting and geo-
    chemical characteristics of the Parbati valley geothermal field, India.
    - In: Proc. of the 2nd UN Symp. on the Dev. and Use of Geothermal
    Resources, San Francisco, California. - US Gov. Printing Office,
    Washington, DC 20402, USA, 1: 329-338.
Craig, H., 1961: Isotopic variations in meteoric water. - Science 133:
    1702.
— , 1963: The isotopic geochemistry of water and carbon in geothermal
    areas. - In: E. Tongiorgi (Ed.), Nuclear Geology on Geothermal Areas.
    - Spoleto meeting, Laboratory of Nuclear Geology, Pisa, Italy, 17-53.
Dansgaard, W., 1964: Stable isotopes in precipitation. - Tellus 16: 436-
    468.
Friedman, I., Redfield, A.C., Schoen, B., and Harris, J., 1964: The
    variation in the deuterium content of natural water in the hydrologic
    cycle. - Rev. Geophysics 2: 177-224.
Gupta, M.L., 1974: Geothermal resources of some Himalayan hot spring
    areas. - Himalayan Geology 4: 492-515.
Gupta, M.L. & Sukhija, B.S., 1974: Preliminary studies of some geothermal
    areas in India. - Geothermics 3: 105-112.
Gupta, M.L., Narain, Hari & Gaur, V.K., 1976: Geothermal provinces of
    India as indicated by studies of thermal springs, terrestrial heat
    flow, and other parameters. - In: Proc. of the 2nd UN Symp. on the Dev.
    and Use of Geothermal Resources, San Francisco, California. - US Gov.
    Printing Office, Washington, DC 20402, USA, 1: 387-396.
Gupta, M.L., Saxena, V.K. & Sukhija, B.S., 1976: An analysis of the hot
    spring activity of the Manikaran area, Himachal Pradesh, India, by
    geochemical studies and tritium concentration of spring water. - In:
    Proc. of the 2nd UN Symp. on the Dev. and Use of Geothermal Resources,
    San Francisco, California. - US Gov. Printing Office, Washington,
    DC 20402, USA, 1: 741-744.
Gupta, M.L., Rakesh Kumar & Singh, S.B., 1979: Present knowledge of Parbati
    Parbati valley geothermal field. - In: M.S. Bhalla & M.L. Gupta (Eds.),
    Proc. of the seminar on the Dev. and Utilization of Geothermal
    Engineering Resources, Hyderabad, Section III: 83-99.
Krishnaswamy, V.S., 1976: A review of Indian geothermal provinces and
    their potential for energy utilization. - In: Proc. of the 2nd UN
    Symp. on the Dev. and Use of Geothermal Resources, San Francisco,
    California. - US Gov. Printing Office, Washington, DC 20402, USA, 1:
    143-156.
Kumar, B., Athavale, R.N. & Sahay, K.S.N., 1980: Use of stable isotope
    method in hydrological investigations with special reference to studies
    in Lower Maner Basin, Andhra Pradesh. - Paper presented at the workshop
    on "Nuclear Techniques in Hydrology", held at N.G.R.I., Hyderabad,
    19-21 March, 1980.
Ravi Shankar, Padhi, R.N., Arora, C.L., Gyan Prakash, Thussu, J.L. & Dua,
    K.J.S., 1976: Geothermal exploration of the Puga and Chumathang geo-
    thermal fields, Ladakh, India. - In: Proc. of the 2nd UN Symp. on the
    Dev. and Use of Geothermal Resources, San Francisco, California. -
    US Gov. Printing Office, Washington, DC 20402, USA, 1: 245-253.

From exploration to production, the geothermal case – an insight to the first R&D programme in the European Community

Pierre Ungemach

with 11 figures and 1 table

Ungemach, P., 1982: From exploration to production, the geothermal case - an insight to the first R & D programme in the European Community. - Geothermics and geothermal energy, eds. V. Čermák & R. Haenel, E. Schweizerbart'sche Verlagsbuchhandlung, Stuttgart: 219-240.

Abstract: An extensive R & D programme in geothermal energy carried out over the past four years by the Commission of the European Communities (CEC) adressed a variety of topic areas relevant to resource exploration, reservoir physics, exploitation of high and low enthalpy sources and to the engineering of man made geothermal reservoirs (Hot Dry Rock, HDR).

Main issues are reviewed as regards exploration strategies, geo-chemistry and geophysics, drilling, well completion and production problems, exploitation constraints and the fracturing of hot, dry and near impervious basement rocks.

Author's address: Directorate General for Research, Science and Education Commission of the European Communities, Rue de la Loi, 200, B-1049 Brussels, Belgium

Introduction

The increasingly harsh energy situation, characterized by a dramatically unbalanced energy supply, faced by the European Community (EC) has indeed been a strong stimulus in launching, in 1975, the first programme in energy common to the nine member states (Belgium, Denmark, Federal Republic of Germany, France, Ireland, Italy, Luxembourg, The Netherlands, United Kingdom). Within the wide spectrum of alternative energy sources geo-thermal energy is worth consideration. Although countries such as Italy and France had long been involved in geothermal matters, either in power generation from dry steam (Larderello field) or in direct use of low enthalpy aquifers (Melun l'Almont space heating doublet), geothermal energy was overlooked and regarded more or less as an exotic curiosity of marginal impact. Today, far sighted scientific approaches are required to overcome the energy price crisis and geothermal energy reenters the energy scene as a respectable partner. A need is felt towards inventorying EC geothermal resources, adding new reserves to existing figures, and ex-ploiting these indigenous resources wherever and whenever appropriate.
　　As a consequence the first R & D programme was promoted with the aim of fulfilling four objectives, respectively:
- acquisition of basic data (subsurface temperatures, heat flow, reservoir parameters),
- improvement of exploration methods, specially geochemistry and geo-physics directed towards specific geothermal targets,

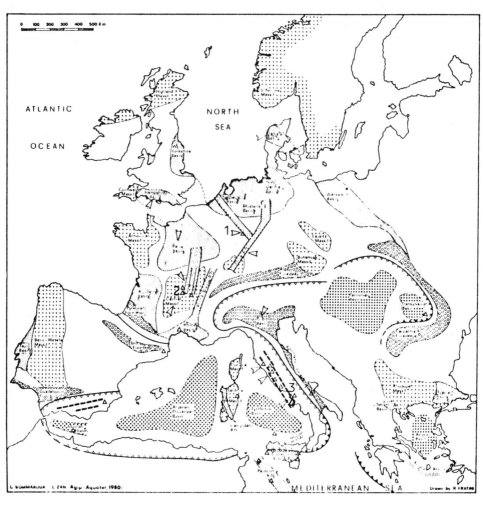

Fig. 1. Main European geodynamic features in relation to geothermal areas (after Sommaruga). 1 Eifel - 2 Mont Dore - 3 Somma Vesuvius - 4 Campi Flegrei.

- development of the models, the experimental procedures and the techno-
  logies required for assessing, producing and managing steam and hot
  water sources,
- feasibility assessment of the Hot Dry Rock (HDR) concept of energy
  mining.

The EC geothermal environment - an outline

The key to the geological structure of Europe and consequently to the
understanding of the origin, location and characteristics of its geothermal resources is given by the geodynamics of the Eurasian plate. In
Fig. 1, which outlines the main geodynamic features of Western Europe
(Sommaruga 1980), can be seen the variety of its geothermal settings which
covers actually the whole scope with the exception of oceanic crust. It
can also be noticed that Western Europe at large is an area of old and
relatively rigid continental crust characterized by crystalline massifs,
intracratonic and foredeep basins and continental rifting. A younger crust
stretches over the mediterranean area at the limit of the Eurasian and
African plates with typical features such as subduction, marginal basins,
island arcs and extensional horst and graben systems. An important
consequence is that the majority of the European States is faced solely
with the low enthalpy outlook, high enthalpy sources being limited to
Central and Southern Italy and to Eastern Greece. Another fact to bear in
mind is the often remote - marine and insular - location of high enthalpy
field.
    Low grade heat can prove to be a prolific and dependable resource in
foredeep and intracratonic basins wherever they develop regional aquifer
units. Such is the case of the Dogger carbonate reservoir in the central
part of the Paris Basin. Here coexist high productive capacities, easy
well completion and a reliable district heating market. Triassic clastic
deposits often pose the problem of random porosity patterns also noticed
in karstified Jurassic limestone. Foredeep basins such as the Northern
Pyreneean foothills, the Molasse Basin in Southern Germany, the Po valley
which develop in a generally subsident context, with filling by sediments
flushed by fresh waters, are cooler than normal. There are, however,
striking exceptions in the Po valley. At Ferrara for example a huge horst
of fractured limestone favours hydrothermal convection and causes
formation temperatures to rise up to 90 °C at a depth of 1 km. Geopressure
can be another distinct feature of the Po valley also encountered in the
North Sea which incidentally is an area of generally high temperature.
High pressures and temperatures are found at great depths (6000 m in
Malossa, Po valley) with dissolved methane in commercial quantities. Here,
however, geopressured horizons exhibit too random porosity patterns to
enable commercial development.
    Continental rifting and central, recent but extinct volcanism are
settings intermediate between low and high enthalpy resources with complex
reservoir and heat source conditions and delicate exploration methodologies. The Rhine Graben is a typical illustration of a two stage rifting
sequence: (a) placing of an extensional valley (mid Eocene - early Miocene)
followed by (b) a reactivation of rifting, modelling the valley into a
sinistral shear belt. There is therefore a combined strike-slip and
distensive transverse faulting system (Illies et al. 1980). In the central
part of the Graben, open shear controlled basement fractures are likely to
favour hydrothermal convection by conveying meteoric waters, infiltrated
in the border massifs, upwards to the buffer reservoir constituted by

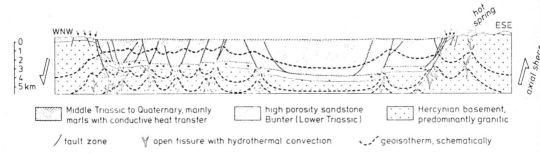

Fig. 2. Generalized cross-section through the Rhine graben South of the
City of Karlsruhe (after Illies et al.).
In the basement, hydrothermal convection is observed in shear controlled
fissures. In the Bunter an additional horizontal circulation evolves, which
is in connection with the recharge areas on the graben flanks. The fine-
grained sediment-fill of the graben conditions a more conductive geothermal
behaviour.

Triassic and/or Jurassic porous sediments (Fig. 2). Locally, the picture
becomes more complicated as to tectonics (characterization and location of
active faults), sedimentology, diagenesis and porosity patterns.
    Magmatism better describes high enthalpy systems than the more re-
strictive volcanic label. As to the heat source - young and shallow hot
bodies - three types of plioquaternary magmatism are distinguished in the
EC, as depicted by Fig. 3, anatectic magmatism, fissural and central
volcanism, fissural volcanism being ruled out with respect to high
enthalpy occurrence. Anatectic magmatism deals with the well known,
singular, geothermal province of Central Tuscany, with the Larderello
prototype of anatectic fusion.
    Recent and active plioquaternary volcanism is extensively studied from
a methodological point of view because of its promising geothermal outcome
and the limited exploration experience available to date. A major objective
of the research was to characterize, by volcanological and petrological
and sometimes geochemical and geophysical methods, the size, shape, depth
and temperature of the respective magma chambers and so define heat supply
mechanism and geothermal expectation. Surveys have been carried out in the
Campi Flegrei, Somma Vesuvius, Latera, Mont Dore and Eifel.
    No investigations have yet been undertaken in the Eolian Island Arc, a
zone of subduction magmatism with promising indices - steam was found in
1953 - and perhaps the highest potential of the Community.

Exploration methodologies

Regardless of low enthalpy prospecting which makes use of oil and ground
water exploration methods, initial hopes based on the development of geophysical techniques designed specifically for high enthalpy prospecting
were illusory. Unfortunately, apart from direct assessment, e.g. drilling,
no surface method exists which is capable of mapping the isotherms of the
subsoil. As a matter of fact, in most instances, exotic geophysical
exploration methods such as seismic noise, magnetotellurics (MT) and SP
did not exhibit very convincing results especially when used individually
with no back-up by conventional exploration tools. These methods have in
common (a) natural sources and non reproduceable data (MT and seismic
noise), (b) limited background knowledge on source processes (SP and low

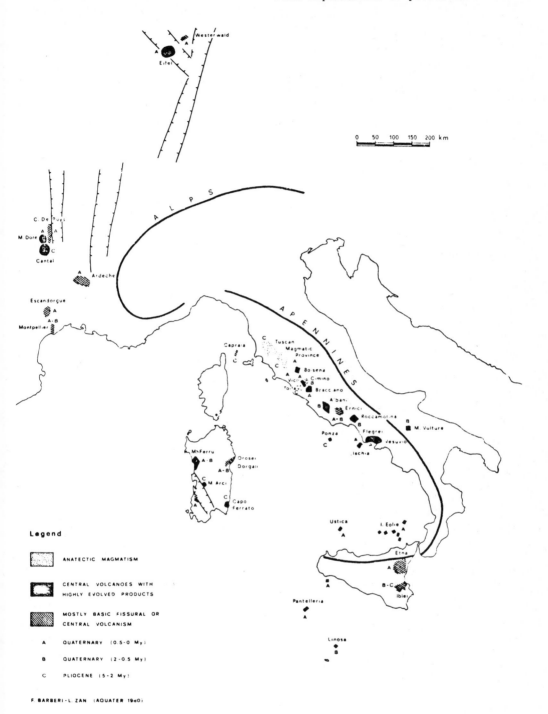

Fig. 3. Plio Quaternary Magmatism in the European Community (after Barberi & Zan Aquater). (From 5 My to present).

frequency telluric currents) and (c) restrictive interpretation models. Channelling, near surface conductive layers, 3D structures often impair the interpretation if not the validity of the source model itself, in MT prospecting. Very often the MT method attributes conductive anomalies to deep seated horizons when in many cases it is caused by a near surface conductive layer (Mosnier 1980 and Berktold 1980). Therefore, more research is needed to appraise the real potential of these methods and a great deal of critical sense is to be exercised in interpreting results (Ungemach 1980).

Redundancy being always welcome in geophysics, it is not surprising that integrated exploration surveys using several methods in conjunction have produced the best results. Not only did they address well defined exploration objectives but they enabled a critical assessment of exploration methodologies. Two case studies will illustrate the topic. The first is the Mont Dore project which combined conventional and novel geophysical methods and resulted in the siting of a wildcat. The second made use of strictly conventional geophysics and led to the discovery of a commercial resource in the Phlegreaen fields.

Mont Dore Survey - Selection of drilling targets

In Mont Dore, area of central, recent but extinct volcanism the aims assigned to integrated reconnaissance were (a) to assess the volcano-tectonic model, (b) to relate it to the occurrence of potential geothermal reservoirs and, (c) to locate a target for a geothermal wildcat (Varet et al. 1980).

The first objective of geophysical surveys was essentially geometric and structural. A strategic campaign based on gravity, airborne magnetics and explosion seismology helped in defining the caldera geometry (contours and depth) not evident from topography. To simplify, the caldera corresponds to a gravity low reflecting a filling by light pyroclastic products. The shallow H field reduced to the pole reveals a series of peripheral magnetic anomalies arranged around a circular structure. This could correspond to basal flows surrounding a caldera (Fig. 4). The deep field reduced to the pole shows a large magnetized body rooted below the Sancy dome, a post caldera event. Processing of gravimetric and magnetic data also indicated discontinuities attributed to the tectonic trends depicted in Fig. 4. Experimental seismology evidenced higher attenuations of P waves in the first two kilometers of the overburden, interpreted as a caldera fill and a fractured basement. This technique deserves a comment owing to its novelty and to its prospecting potentialities in mountainous areas. Detection in the attenuation spectra of geothermal abnormal bodies, say one kilometer in size located at depths lower than five kilometers, implies higher resolution e.g. shorter wave lengths than those obtained with classical seismology and teleseismic events. Hirn et al. (1980) designed a method known as seismic transmission which provides higher frequency signals (2-20 Hz) from explosive sources and Moho reflected waves at critical incidence. In Massif Central, Moho depth is known and is fairly uniform. Medium charge explosive fired in different azimuths and P_MP waves achieve a kind of bottom illumination of the anomalous body. Delays in time arrivals are monitored on a dense array of three-component seismometers (up to 150 in Mont Dore). Such waves, not being first arrivals, special expertise is needed in matching P_MP arrivals times before a consistent criterion be available. The method was successfully tested in Mont Dore. Here, 3D inversion (Aki et al. 1977 and

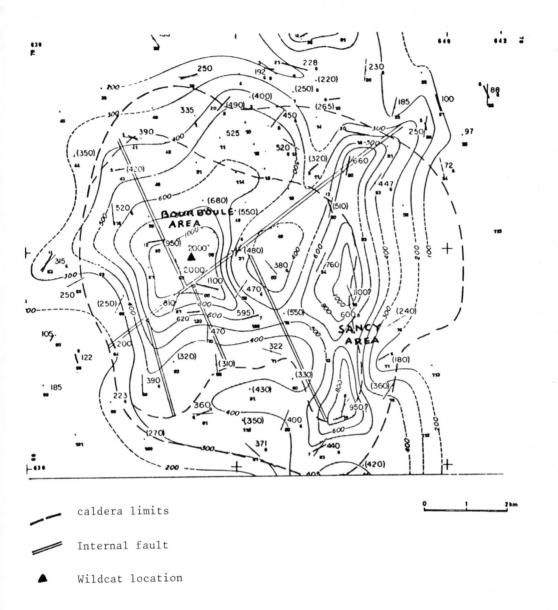

Fig. 4. Longitudinal conductivities in the Mont Dore caldera (after Varet et al.).

Nercessian 1980) enabled to identify a low velocity volume which happened to coincide with the caldera contours derived by other methods.

Tactical surveys undertaken in a second phase inside the caldera combined conventional (geoelectrics, reflection and refraction seismics) and exotic (magneto tellurics (MT) seismic noise, SP) prospecting techniques. In particular low frequency MT's spotted two highly conductive areas (Fig. 4) a pronounced circular one to the North-West and two twin lower anomalies eastwards. Here, the effect of the border fault is manifest with elongated anomalies and high polarisation. Refraction and reflection seismics indicated both fractured basement rocks and deep reflectors in the North-Western area, already inferred from explosion seismology.

Silica and alkaline geothermometers (SiO_2, Na/K, Na-K-Ca and Na/Li) applied to the spring waters flowing along the valley which closes (fault) to the North the zone of first interest, gave comparable results with no indication of mixing whatsoever. Source temperatures range from 140 to 170 $^{\circ}$C depending upon spring location. This was another argument in favour of this area with respect to a first exploratory borehole (see Fig. 4). So, field reconnaissance could solve the initial dilemna on whether the spring area or the youngest and presumably hottest Sancy down should be selected first on the basis of reservoir oriented indices.

Phlegreaen fields - Discovery of a commercial reservoir

The Phlegreaen fields are a different context than Mont Dore. It is an active volcanic area with historic episodes dating back to 400 years ago and near surface hot bodies. Quaternary alkaline volcanism is widespread over the Campanian plain, chiefly a large graben filled with mesozoic carbonate sediments dipping from the Apennines towards the sea along a NE-SW (anti-apennine) trend.

Volcanic activity can be divided into three main sequences from 70,000 years ago to recent times. The Mofete hill located Westwards (Fig. 5) was placed as a part of a volcano of the second phase, some 10,000 years ago, then partially destroyed as a consequence of volcano-tectonic events (Barberi et al. 1980).

Structurally the Mofete hill belongs to the rim of the vast sinking caldera materialised by the Pozzuoli Gulf. It is an uplifted fault block system higher than the surrounding compartments.

The prime objective of the survey was to derive a comprehensive conceptual model of the area in relation to the presence of shallow hot bodies and an intense fracturation, partially masked by a recent volcanic cover, favouring either solid (magma filled vents) or liquid (hydrothermal) convection.

Conventional geophysical methods were first applied in order to portray its main structural and tectonic features (AGIP 1980).

(a) Gravity surveys. The Bouguer anomaly reflects a large central low, subcircular in shape, coinciding with the Pozzuoli caldera and surrounded by a ring of small local positive anomalies (Fig. 5). High pass filtering enabled to produce a map of the residuals which pinpoints a succession of gravity highs close to the gulf interrupted by discontinuities trending radially. Positive residuals indicate a density contrast of locally denser rocks placed in a light environment which can be attributed either to heavy lavas or to self-sealed cap rocks. Discontinuities correspond most likely to fault and fissures which act as vents and paths for magma or hydrothermal fluid rises.

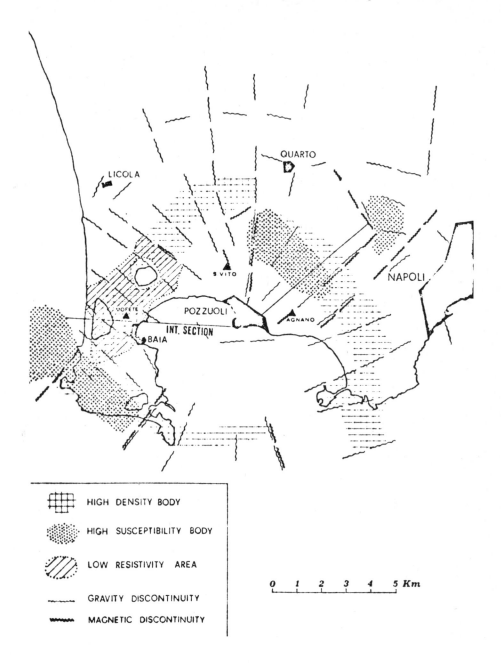

Fig. 5. Geophysical composite interpretive map of Campi Flegrei (after AGIP).

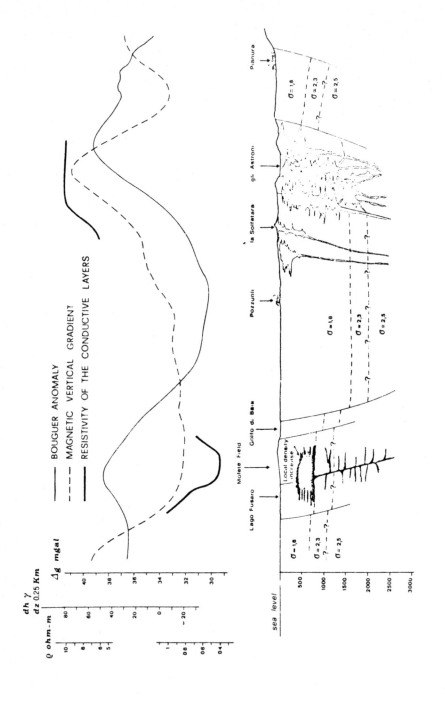

Fig. 6. Interpretive cross-section of the Pozzuoli caldera (after AGIP).

(b) Airborne magnetics. The total H field was processed in order to obtain, after filtering of higher frequencies, a map of the vertical gradient reduced to the pole in view of isolating the magnetic contribution of deep seated sources. The interpretation, as it often occurs, is somewhat ambiguous as to magnetization of causative bodies. In brief, positive anomalies could correspond to strongly magnetized materials ascending through fractures. Such lavas have in fact higher susceptibilities than those of the pyroclastic, mainly tufaceous, cover. Negative anomalies could relate either to demagnetisation by a hydrothermal flush (magnetite is transformed into iron hydroxyde) or to the lack of lavas of sufficient size. Here also radial discontinuities appear which strongly suggest an array of typical pericalderic vents and ring dykes. Of interest is the correlation of gravity residuals and of magnetic anomalies. Wherever magnetic and gravity highs are in phase as to the East of the caldera (Pozzuoli solfatare) it can be assumed that lava filled vents and fractures are present; whereas in the Western part (Mofete block) reverse trends (gravity residual high and magnetic low) could reveal a self-sealing process associated with hydrothermal convection through fractures in distensive state (see interpretive sketch in Fig. 6).

(c) Geoelectrics. In order to discriminate solid and liquid filled compartments, a resistivity survey (VES Schlumberger array) was undertaken which clearly indicates a conductive horizon (0.5 to 1 Ωm at a depth ranging from 600 to 1000 m) in the Mofete area as opposed to the resistive Eastern part. Low resistivities could correspond to a conductive selfsealed cap rock or to a hot geothermal brine or to both.

Hydrothermal leaks could be spotted by a detailed hydrogeochemical survey of shallow watertable aquifers (Cioppi et al. 1980). In particular, the mapping of H_3BO_3, CO_2 and NH_3 anomalies could be related to the presence of vertical interactions occurring through active fracture systems connecting deep and shallow horizons. A low H_2/CH_4 ratio combined with a high nitrogen content was selected as a self-sealing indicator (no CO_2 rise) (Cioppi et al. 1980).

In conclusion, and although some aspects of deep magnetic anomalies remain ambiguous, this model was validated by the three productive wells recently drilled in the Mofete area which indicate the presence of a commercial reservoir (AGIP 1980).

Low temperature resources - development of suitable surface processes

The increase of the geothermal reserve of the EC is strongly dependent on exploration drilling and also on the development of its large resources of low grade heat. Drilling deeper for searching higher temperatures is one route. But, provided compaction does not drastically reduce porosity and yields, the higher well costs require substantial heat loads in the close vicinity of the well, say at least 3000 equivalent dwellings, each 185 m^3 in volume, at a distance lower than 5 km with a load factor of 70 % for a 3000 m deep well-pair (the socalled geothermal doublet). This indeed restricts development to a limited number of privileged sites e.g. large urban areas. Hence, higher temperatures and deeper wells may not be the answer unless, paraphrasing a famous humorist, cities be removed to the countryside.

Another approach would consist of tapping shallower and consequently cooler but also fresher aquifers that could economically supply reduced heat loads. This would undoubtedly add some flexibility to the economics -

one shallow well instead of two deep ones, and the possibility of a joint fresh water and heat supply – and widen the usage, should the appropriate low temperature process exist. Two ways are possible, either incorporate a heat pump to the heating circuit or design a heating mode compatible with low water inlet temperatures in the 40 to 60 °C range. Both have been tested and are currently experienced and monitored.

Finally, the most promising alternative proved to be a low temperature air-water convector utilising a natural convection principle on the air face (Bertin 1980). The driving idea behind the concept is simple. Given the power of a convector expressed as:

$$W = \alpha S \, \Delta T$$

where
α = heat exchange coefficient
S = heat exchange area
$\Delta T = (T_w - T_a)_{log}$

T_a, T_w = air and water temperatures

the decrease of ΔT must be compensated by an improvement of either the heat exchange coefficient or the heat exchange area or both. In fact, for obvious geometric reasons most of the effort is to be placed on the increase of α. Let us now consider a conventional commercial convector currently operating at a water inlet temperature of 80 °C. The arithmetic temperature difference is of 60 °C (room temperature 19 °C), whereas for a water inlet temperature of 50 °C it reduces to 30 °C and becomes highly sensitive to the value of the outlet temperature. Thus, when moving to logarithmic ΔT's it is a gain in performance of almost 3 which is to be achieved over conventional devices. Such improvements could be obtained by designing a prototype which optimizes the heat transfer by (a) an assembly of layers of finned tubes, in which water circulates at low velocities, equivalent to a counter-current sweeping (b) a natural convection enhanced by a chimney effect, realized by a vertical duct; its height is almost equal to that of the room in order to compensate the reduction in draft power caused by the low temperature gradients on the air side. As a matter of fact this appeared to be a serious advantage with respect to the homogeneisation of room temperature. The performances of the qualified prototype are summarized in Fig. 7. Of interest are

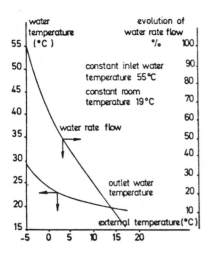

Fig. 7. Air water-low temperature convector. Inlet-outlet temperatures versus ambient temperature. Room temperature 19 °C (after Bertin).

(a) the outlet temperature of 29 °C achieved by an ambient temperature of
-5 °C (geothermal and room temperatures of 55 and 19 °C respectively) and
(b) the thermal output of 750 W_t for the afore-mentioned temperatures.
Research continues in this field, which adresses the monitoring in real
conditions over two winters, and the design of a forced convection prototype capable of delivering 750 W_t with geothermal waters at 40 °C, under
a reduced volume. Already, the cost of substitution of the convector to
existing systems proves competitive. A rough estimate indicates elsewhere
a twofold increase of the geothermal reserve of the EC by incorporating
fluids at temperatures comprised between 40 and 60 °C.

Hot dry rock

What is HDR all about?

The heat stored in the uppermost part of the Earth crust exceeds by
several thousands of orders of magnitude the resources of all known fossil
fuels. Given an average rock density of 2500 kg/km^3 and a specific heat of
1000 J/kg/°C the energy stored in one km^3 of rock amounts to 25.10^{16} J/°C.
This *resource base* has been estimated by the Los Alamos Scientific
Laboratory (LASL), over the entire mainland USA, down to a depth of 10 km,
for temperatures in excess of 150 °C and whatever the basement rocks, at
4 billions Megawatts (thermal) centuries as compared to 400,000 and 4
millions of hydrocarbon and coal origins respectively. The resource base
relevant to hot water and steam sources has been estimated by the same
authors at 540,000 MW_t centuries considering no resupply of the heat in
place neither by conduction nor solid or liquid convection.
  This means that, although not commercial, a considerable resource base
exists which could become a reserve should a feasible terrestrial heat
mining process be designed at some future time. It would make it possible
to extract energy from hot, dry and impervious basement rocks over say
80 % of emerged lands e.g. almost anywhere whereas Geology bounds the
development of natural geothermal energy sources to privileged sites.
  The site free availability of such an enormous potential resource is
the driving mechanism of the HDR concept and of its dominating engineering
philosophy. It is not surprising, therefore, that the concept became
quickly popular among energy planners everywhere. It would raise, give or
take a few details on the recovery process, the potential contribution of
geothermal energy from 5 % of the energy demand to nearly 20 %. From its
present marginal position geothermal energy would become a more respectable
partner with the HDR process even constituing for some people the only
long term future for this energy source.
  Now, let us adopt mining arguments. When cooled down by 100 °C a one
km^3 of rock volume stores an energy of 25.10^{17} J. Assuming further the
temperature drawdown remains constant over a 20 year life span and a
recovery factor of 1 % - about one half the figure achieved by a natural
steam reservoir - would lead to a power output of 4 MW_t.

The HDR concept

Energy recovery from hot, dry and impervious rocks requires the engineering
of a downhole heat exchanger. This implies that a heat carrier fluid be
circulated over a large heat exchange area, between a pair of wells
connected through a low hydraulic impedance, e.g. the creation of an

artificial hydrothermal convection reservoir. The requirements are three-fold: (a) to design a stimulation process capable of generating a re-servoir of sufficient capacity, (b) to circulate the vector fluid at large mass flow rates in a closed loop pressurized system and (c) to achieve a heat transfer area allowing to sweep the heat in place in commercial quantities.

Typical figures thought to secure a viable operation of a HDR doublet (see Fig. 8) are (CEC 1980):
- heat transfer area = 5 km^2
- stimulated bulk volume = 5-10 km^3
- recovery factor = 1-2 %
- circulated mass flow rate = 100 kg/s
- temperature difference = 120 °C
- maximum well head pressure = 100 bars
- lifetime = 20 years
- minimum installed capacity = 50 MW$_t$.

Among the three successful field tests – all achieved linking and all re-opened existing fractures – carried out in the granite at depths ranging from 150 to 300 m in France (Mayet de Montagne – Massif Central), Germany (Falkenberg – Bavaria) and in the United Kingdom (Rosemanowes – Cornwall) the latter deserves a special comment. There is geological evidence that

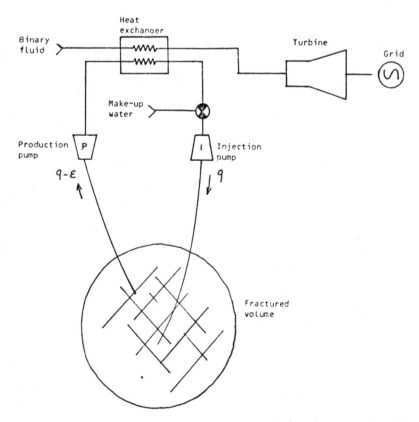

Fig. 8. The Hot Dry Rock concept. A typical HDR doublet.

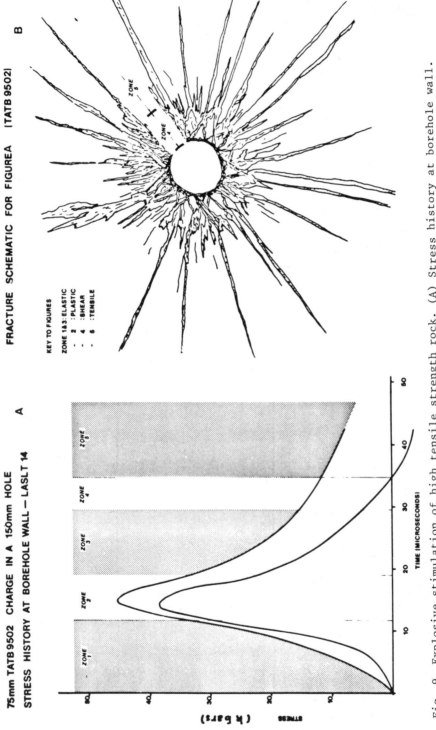

Fig. 9. Explosive stimulation of high tensile strength rock. (A) Stress history at borehole wall. (B) Fracture schematic (after Batchelor and LASL).

basement rocks are naturally fractured. But, in Cornwall, the evidence is more visual perhaps than anywhere else because of the numerous quarries and mine galleries, some 1000 m deep, existing in the Cornish batholyth. Here, the intense rock jointing was induced by the cooling of the pluton. This primitive feature added to the high tensile strength of the granite led to a different approach to fracturing.

Jointing causes rock strength to be discontinuous and the combination of a strong basement rock fabric and of planes of weakness is likely to be the major controlling factor in the initiation and propagation of induced fractures. Hence, a multiple connection between wells seems possible and the concept of linking two or more wells through a series of reopened joints replaces the model of a single planar fracture, oriented perpendicular to the least principal stress, that prevails in unbroken rocks (CEC 1980).

The key idea behind explosive pretreatment is to beat the local stress accumulation at the well bore by propagating a radial array of cracks which, further enhanced by hydraulic pressure, will access the natural joint network.

Provided the explosive is fired below the plastic limit, explosion modelling shows that shock wave induced stresses can build up such an array, which propagates concentric to the well at a distance of 8 to 10 times the borehole radius (see Fig. 9). This process creates a self-propped, strongly negative skin zone, which clearly increases well injectivity, reduces breakdown pressures and bridges the difference between initiation and propagation pressures therefore easing the access to the joint network. Further hydraulic pressurization cycles can therefore more reliably propagate fractures to intersect and connect existing joints.

The process operated satisfactorily at a depth of 300 m as reported by Batchelor et al. (1980). Although the blast was not fired at design figures the experiment achieved the linking of three wells through at least three distinct fractures with bottom hole distances varying from 9 to 20 metres. The interpretation of isotope tracer tests, though somewhat ambiguous, could identify up to 7 connecting flow paths. The loop underwent a three week back pressure test at a rate of 10 kg/s under a WHP of 63 bars, just below refrac pressure, and exhibited a 99 % flow recovery ratio. The explosives used in the detonator, booster and main charge assembly are declassified military explosives which display a poor oxygen balance and stability at high temperatures. TATB (Tri amino trinitro-benzene $C_6H_6N_6O_6$) the main charge, is even reported to be cleared from a hole by drilling. The improvement brought by explosive initiation is visualized in Fig. 10 (a, b and c). Fig. 10a indicates that the first test failed in achieving fracturing at 9.2 bars, in the absence of explosive pretreatment. A second test reached breakdown pressure at 8.4 bars, the curve suggesting the reopening of a preexisting natural fracture. Finally, the build-up curve following the explosive stimulation of the first tested horizon shows that pressure recovers the refrac pressure of 8.4 bars noticed (test no 2) in the case of a natural discontinuity. The smooth shape of the recovery curve and a sonic log inspection of the hole tend to indicate a progressive opening of fissures around the well bore. Fig. 10b shows the improvement of the injectivity index following the explosion (test RT 240 against test RT 220) and Fig. 10c the nearly instantaneous closed loop circulation (99 % recovery) established by the explosive initiated reservoir.

Other field experiments emphasized fracture mapping techniques (Cornet 1980 and Kappelmeyer 1980), hydrofracturing in situ stress measurements (Cornet 1980) and hydrothermal testing (Hosanski 1980). In France and

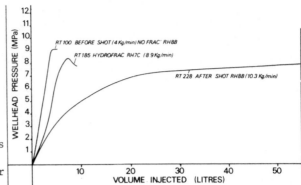

Fig. 10a. Rosemanowes quarry HDR site. Comparison of hydraulic stimulation results with and without explosive pretreatment (after Batchelor et al.).

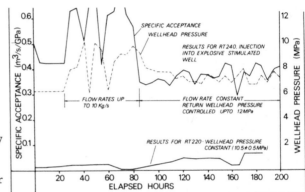

Fig. 10b. Rosemanowes quarry HDR site. Comparison of specific acceptance during circulation (after Batchelor et al.).

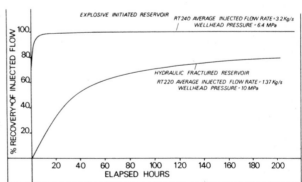

Fig. 10c. Rosemanowes quarry HDR site. Comparison of percentage recovery during circulation tests (after Batchelor et al.).

Germany, the field procedure can be summarized as follows. A first borehole is drilled at some 200 m and hydrofractured. Then, based on the mapping of the hypocenters of the during and post fracturing acoustic and microseismic emissions a second well is drilled to intersect the fracture and establish a lower hole connection. In Bavaria this latter part was delicate owing to a nearly vertical fracture and a non directional drilling facility. At Mayet de Montagne the thermal conductivity and the specific heat of the host rock were identified in situ and a temperature breakthrough was detected on the production hole after a 46 day injection, in a 30 m distant well.

Future research implications of the HDR concept

There is at least one objection to the significance of the aforementioned tests. They have been conducted under a state of stress not representative of the stresses (and temperatures) existing at ultimate HDR depths (5 to 7 km in areas of normal geothermal gradient against 300 m in present conditions).

Another uncertainty is the efficiency of explosive pretreatment at target depths. The absolute difference between initiation and propagating fluid pressure being in our case equal to:

$$\Delta p = \lambda p z + T$$

where: T = rock tensile strength
          z = depth
          p = rock density
          λ = ratio between minimum horizontal and vertical rock stress

the advantage in terms of pressure gradients $\Delta p/z$ decreases at greater depth with T and also λ (isotropic behaviour) unless tectonic stresses decide otherwise (CEC 1980 and Geerstma 1980). Here again, experience is needed to correctly appraise the effect on an anisotropic stress field in a deepseated basement rock.

A third constraint is the unavailability of hardware and routine procedures at elevated bottom hole temperatures. No such equipment and practice exist yet, on commercial basis, in particular in the field of logging, open hole packers and directional drilling steering tools, vital to HDR engineering.

On the other hand, LASL experiments have shown that casing, threads and, moreover, cement bonds can hardly resist successive cooling sequences and related thermal shocks applied to the injection well bore. Prestressed casing and tensile pumping of cement could be a remedy but it requires field validation.

Propping of the fracture(s) is another prerequisite to the successful operation of a HDR loop. Solutions adopted in the oil industry cannot be transferred straight forward to HDR's. So, self-propping of fractures must be the rule in a closed loop circulation, to avoid the pumping of propping agents which would have an adverse effect. Leaching could be an attractive solution. Not only does it markedly improve fracture conductivity, as displayed by Fig. 11a, as a consequence of the circulation of a 10 % hydro-alcoholic solution of sodium hydroxyde, but it warrants, by silica dissolution in a coarse granite fabric, a propping of the fracture. The kinetics of pH recovery towards neutrality, are essential to avoid secondary precipitations. The optimum leaching sequence was found to be that described in Fig. 11b which maintains sample permeability at its

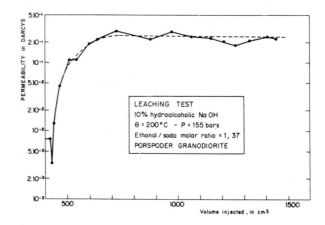

Fig. 11a. Chemical leaching experiments. Permeability increase of a granodiorite sample versus injected volume of NaOH (after Sarda et al.).

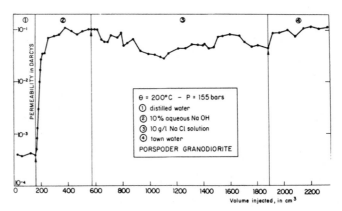

Fig. 11b. Chemical leaching experiments. Optimum leaching sequence (after Sarda et al.).

optimum level (Sarda et al. 1980).

But more important, however, is system lifetime. Intuitively one can imagine that an optimum HDR operation requires a compromise between circulated mass flows, which define installed capacity and system rating, and the heat exchange area which in turn fixes lifetime. For the sake of illustration let us simplify the mathematics of heat transfer by assuming the temperature transients can be averaged over the entire frac area by a one dimensional diffusion process. Hence, standard heat conduction theory yields the following temperature response:

$$\frac{T - T_{min}}{T_i - T_{min}} = \mathrm{erf}\ (\frac{s}{q}\ \Theta) = T_D$$

where: T = temperature at time t

T_i = mean initial rock temperature

T_{min} = water reinjection temperature

S   = heat exchange area

q   = water mass flow

Θ = $(\lambda r p_r C_r / t)^{\frac{1}{2}} / C_w$

C_w = water specific heat

C_r = rock specific heat

p_r = rock density

λ_r = rock thermal conductivity

erf = error function.

Calculations performed for water - C_w = 4200 J/kgK - and an "average rock" - C_r = 1000 J/kgK, λ_r = 3 W/mK and p_r = 2500 kg/m^3 - yield the temperature decay transients displayed in Tab. 1 with S/q as a parameter. Setting the lifetime to 0.9 ($T_i - T_{min}$), which is less restrictive and more realistic than the strict breakthrough time, enables to evaluate the minimum heat exchange area. Given the fact that an installed capacity of 50 MW$_t$, expressed as $P(MW_t) = 4.1810^{-3}q$ ($T_i - T_{min}$), is needed at well head, a mass flow of 100 kg/s under a temperature difference of 120 °C is necessary. This means that an effective heat exchange area of 5 km^2 will be required to secure a lifetime of 25 years and therefore concept viability. Clearly, channelling is in this respect the greatest fear specially when considering the volumetric multipath connection suggested by preliminary field trials. A number of independent connections should lower loop impedance but it could as well dramatically reduce the heat transfer area. Leaching could aggravate this situation by enhancing the hydraulic conductivity of preferential conduits enlarged by chemical attack (the fracture conductivity, in laminar flow, is proportional to the square width).

Tab. 1. Lifetime of a Hot Dry Rock well pair as a function of mass flow and heat exchange area.

$\dfrac{T-T_{min}}{T_i-T_{min}}$ S/q	10^3	10^4	2.10^4	5.10^4	10^5
0.9	0.01	1	4	25	100
0.5	0.06	6	24	125	600
0.1	1.67	167	668		

Temperature decay time (in years)

Fracture spacing is another implication on lifetime as to thermal interferences. Analytical calculations (Carslaw & Jaeger 1941) show that a set of parallel equidistant fractures can achieve a 90 % recovery of the heat stored in an homogeneous and isotropic slab provided the parameter $\alpha = 4 \lambda_r t/C_r d^2$ is equal to unity (d = fracture spacing, t = time). Adopting the just mentioned values for rock a lifetime of 25 years would require a spacing of 20 meters. A one meter spacing would in turn reduce the lifetime to ten days!

An important remark needs to be introduced here, which is thermal stress cracking studied by Murphy (1979) among others. The thermal contraction caused by the cooling of the rock can initiate cracks once the thermally induced tensile stress exceeds the compressive strength of the rock. This would undoubtedly benefit to the operation of a HDR loop, thermal stress cracking developing, towards the hottest and deepest parts of the rock, proportionally with cooling of the mass.

Acknowledgements. This work forms part of the first and second R & D programmes on alternative energy sources (subprogramme Geothermal Energy) sponsored by the Commission of the European Communities, Directorate General for Research, Science and Education.

The author is indebted to the many programme contributors and in particular to Dr. C. Sommaruga, G. Baron, R. Haenel, J.P. Sarda and A. Batchelor for stimulating discussions on many aspects of geothermal energy research.

References

AGIP S.p.A., 1980: Mofete 1 and 2 wells. Final Report, San Donato Milanese, 30 June, 1980. - EC contract GE 01/79, 23 pp., 15 fig., 9 tab.
Aki, K., Christofferson, A. & Husebye, E.S., 1977: Three-Dimensional Seismic Structure of the Lithosphere. - J. Geophys. Res. 82: 277-296.
Barberi, F., Iannaccone, G., Innocenti, F., Luongo, G., Munziata, C., Pascale, G. & Rapolla, A., 1980: The Evaluation of a Preliminary Geothermal Model for the Phlegreaen Fields Volcanic Area. - In: A.S. Strub & P. Ungemach (Eds.), Advances in European Geothermal Research. - Reidel Pub. Co., Dordrecht, 121-140.
Batchelor, A.S., Pearson, C.M. & Halladay, N.P., 1980: The Enhancement of the Permeability of Granite by Explosive and Hydraulic Fracturing. - In: A.S. Strub & P. Ungemach (Eds.), Advances in European Geothermal Research. - Reidel Pub. Co., Dordrecht, 1009-1031.
Berktold, A., Kemmerle, K. & Neurieder, P., 1980: Magnetotelluric Measurements and Geomagnetic Depth Sounding in the Area of the Urach Geothermal Anomaly. - In: A.S. Strub & P. Ungemach (Eds.), Advances in European Geothermal Research. - Reidel Pub. Co., Dordrecht, 911-920.
Bertin et Cie., 1980: Etude d'Echangeur Air-Eau à Tubes Ailetés Fonctionnant à Bas Reynolds Permettant la Conception de Convecteurs Basse Température pour le Chauffage de Locaux. - Rapport de fin de Recherche. - Contrat CEC no 638-78-9-EG F. 60 p., 8 fig., 1 ann. confidentielle.
Carslaw, H.S. & Jaeger, J.C., 1959: Conduction of Heat in Solids. - 2nd Ed., Clarendon Press, Oxford.
Cioppi, D., Ghelardoni, R., Panci, G., Sommaruga, C. & Verdiani, G., 1980: Demonstration Project. Evaluation of the Mofete High Enthalpy Reservoir. (Phlegreaen Fields). - In: A.S. Strub & P. Ungemach (Eds.), Advances in European Geothermal Research. - Reidel Pub. Co., Dordrecht, 291-302.

Commission of the European Communities (CEC), 1980: The European Hot Dry
    Rock Case. Where to Go Next? An Assessment in Methodology by the
    Working Party on HDR Research and Technology. - Brussels, 18 June 1980.
    (In Press).
Cornet, F.H., 1980: Analysis of Hydraulic Fracture Propagation. A Field
    Experiment. - In: A.S. Strub & P. Ungemach (Eds.), Advances in European
    Geothermal Research. - Reidel Pub. Co., Dordrecht, 1032-1043.
Geerstma, J., 1980: Contribution to the EC Hot Dry Rock Working Party. -
    Comments on the Use at Great Depth of Explosives to Initiate Hydraulic
    Borehole Fracturing. 16 April, 1980. Unpub. Doc.
Haenel, R., 1980: The Subsurface Temperature Atlas of the European
    Communities. - Paper presented at the 7th annual meeting of the
    European Geophysical Society, Budapest, 24-29 August, 1980. (In press).
Hirn, A. & Nercessian, A., 1980: Identification of Three-Dimensional
    Bodies by Moho Reflected Waves. Application to the Mont-Dore Area. -
    In: A.S.Strub & P. Ungemach (Eds.), Advances in European Geothermal
    Research. - Reidel Pub. Co., Dordrecht, 622-633.
Hosanski, J.M., 1980: Contribution à l'étude des Transferts Thermiques en
    Milieu Fissuré. - Thèse Docteur Ingénieur - Université Pierre et Marie
    Curie - Paris VI. 19 sept., 1980.
Illies, J.H. & Greiner, G., 1978: Rhinegraben and the Alpine System. -
    Geol. Soc. Amer. Bull. 89: 770-782.
Illies, J.H. & Hoffers, B., 1980: Neotectonics and the Geothermal
    Anomalies in the Rhine Graben. - In: A.S. Strub & P. Ungemach (Eds.),
    Advances in European Geothermal Research. - Reidel Pub. Co., Dordrecht,
    50-61.
Kappelmeyer, O. & Rummel, F., 1980: Investigations on an Artificially
    Created Frac in a Shallow and Low Permeability Environment. - In:
    A.S. Strub & P. Ungemach (Eds.), Advances in European Geothermal
    Research. - Reidel Pub. Co., Dordrecht, 1044-1053.
Los Alamos Scientific Laboratory (LASL), 1980: Hot Dry Rock Geothermal
    Project Progress Reports for Fiscal Years 1977, 1978 and 1979. - Los
    Alamos, New Mexico, USA.
Mosnier, J., 1980: Utilisation of Magnetic, Electric and Electromagnetic
    Methods in Geothermal Prospecting. Present Limitations and Future
    Prospects. - Paper presented at the 26th International Geological
    Congress, Section S 14.2.1., Paris 16-17 July, 1980. (In press).
Nercessian, A., 1980: Heterogeneité Crustale sous le Massif Volcanique du
    Mont Dore; Essai de Prospection par une Méthode de Sismique Trans-
    mission. - Thèse Doct. 3ème cycle, Univ. Paris VII, 24 Oct. 1980.
Sarda, J.P. & Roque, C., 1980: Permeability Stimulation of Crystalline
    Rocks by Chemical Leaching. - In: A.S. Strub & P. Ungemach (Eds.),
    Advances in European Geothermal Research. - Reidel Pub. Co., Dordrecht,
    977-988.
Sommaruga, C., 1980: High and Low Enthalpy Geothermal Resources Explora-
    tion: Models, Strategics and Realities. - Paper presented at the 26th
    International Geological Congress. Section S 14.2.1., Paris 16-17 July,
    1980. (In press).
Ungemach, P., 1980: Geothermal Resources Exploration in the European
    Community. The Geophysical Case. - Paper given at the Workshop on the
    Geophysics of Geothermal Areas: State of the Art and Future Develop-
    ment. Ettore Majorana International Centre for Scientific Culture.
    Advanced School of Geophysics. Erice - Italy. 21-29 May, 1980. (In press).
Varet, J., Stieltjes, L., Gerard, A. & Fouillac, C., 1980: Prospection
    Geothermique Intégrée dans le Massif du Mont Dore. Synthèse. Rapport
    final de recherche. - Contrat CEC no 489-78-EGF. 38 p., 34 fig., 6 tab.,
    4 ann.

Geothermal exploration in the hot spring area Baden-Schinznach, Switzerland*

Ph. Bodmer, F. Jaffe, L. Rybach, J.F. Schneider, J.P. Tripet, F. Vuataz and D. Werner

with 4 figures and 1 table

Bodmer, Ph., Jaffe, F., Rybach, L., Schneider, J.F., Tripet, J.P., Vuataz, F. & Werner, D., 1982: Geothermal exploration in the hot spring area Baden-Schinznach, Switzerland. - Geothermics and geothermal energy, eds. V. Čermák & R. Haenel, E. Schweizerbart'sche Verlagsbuchhandlung, Stuttgart: 241-248.

Abstract: Detailed geological, geochemical and geophysical investigations have been carried out in the hot spring area Baden - Schinznach, Switzerland, in order to clarify the origin and the geothermal potential of the thermal water system. All existing springs and boreholes have been observed periodically in order to determine the chemical and physical characteristics of the thermal and nonthermal fluids. The application of different chemical geothermometers indicated the reservoir temperature of different water types. The occurrence of warm water with temperatures up to 48 °C and high mineralization up to 4.5 g/l is strongly linked to the intersection of the main Jura overthrust ("Hauptüberschiebung") and a system of subvertical north - south striking faults. Detailed geological and geophysical surveys made it possible to locate 20 shallow and three 70 - 135 m deep drillholes in order to obtain more information about the hydrogeology and the geothermal conditions of the most promising parts of the area under investigation. Geological description, aquifer tests, well logging and water sampling for geochemistry have been performed in those drillholes. One of them was put into commercial production. Computer simulation of the thermal water system together with the interpretation of structural and geochemical data as well as heat flow determinations led to a model of the thermohydraulic conditions in the deep underground and to a yield estimate of the system under study.

Authors' addresses: Ph. Bodmer, L. Rybach, D. Werner, Institute of Geophysics, Swiss Federal Institute of Technology, Zurich, Switzerland; F. Jaffe, F. Vuataz, Department of Mineralogy, University of Geneva, Geneva, Switzerland; J.F. Schneider, J.P. Tripet, Motor Columbus Consulting Engineers Inc., Baden, Switzerland

Introduction

The investigated area belongs to the Tabular and Folded Jura, and lies between the Black Forest massif to the north and the eastern end of the Folded Jura to the south (Fig. 1). Along the tectonic contact between Tabular and Folded Jura (main Jura overthrust = "Hauptüberschiebung") several thermal springs are located, especially at places where this

* Contribution no. 323, Institute of Geophysics ETH Zurich

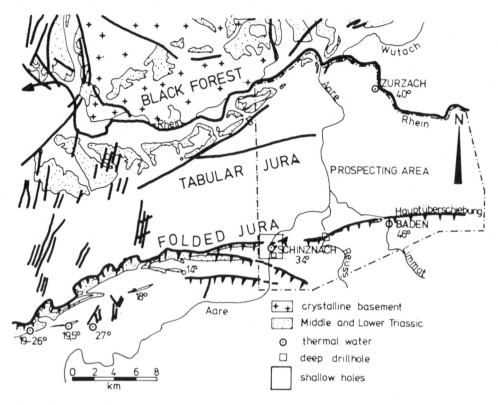

Fig. 1. Tectonic map of the area under study. Note the numerous thermal springs along and south of the "Hauptüberschiebung".

overthrust intersects a subvertical north – south trending fault system. The thermal-water flow at Baden measures 13 l/s at an outlet temperature of 46 °C, while the old thermal spring of Bad-Schinznach delivers 4 l/s at 34 °C. In Zurzach, at the foot of the Black Forest massif, two 430/470 m deep drillholes penetrating the sediments of the Tabular Jura produce together 11 l/s water at 40 °C from the top of crystalline basement (Jaffe et al. 1980).

The geologic column starts with the crystalline basement (Fig. 2) and is covered by Mesozoic (limestones, dolomites, evaporites, marls) and Tertiary sediments (sandstones and marls of the "Molasse"). Two layers in this series are known as warm water producing aquifers: a karstic lime-stone (Triassic Muschelkalk) and the weathered/fractured crystalline base-ment. More detailed information about the general geology can be found in Motor Columbus (1977) and in Schindler (1977).

Based on an early feasibility study (Motor Columbus 1977), a systematic geothermal exploration was started in this area in order to clarify the origin, mechanism and the geothermal potential of the thermal water system. At the same time, several hydrogeological, geochemical and geo-physical methods were tested and appraised in view to elaborate further exploration programs in geothermally promising areas in Switzerland. The estimate of the geothermal potential should enable decision-making on siting geothermal pilot installations for space heating.

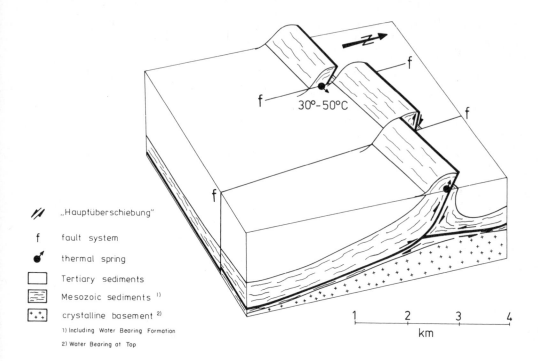

Fig. 2. Schematic section of the overthrusted Folded Jura. Thermal springs issue at places where the "Hauptüberschiebung" is intersected by the system of subvertical faults (f).

Geology and geophysics

The investigations started with a compilation of all available structural, geological and hydrogeological information. Simultaneously all springs were sampled along with temperature and electrical conductivity measurements. Geoelectric, gravimetric and seismic refraction measurements were carried out and supplemented by detailed geological mapping in order to obtain structural information on the zones of uprising thermal water. The geophysical measurements also furnished the thickness of quaternary cover.

To evaluate the geothermal potential of the thermal water system the regional as well as the local temperature field has to be determined. For this purpose, continuous temperature logs were measured (in thermal equilibrium) in 55 existing drillholes within and around the project area and representative rock samples were taken for thermal conductivity determinations. The temperature data were corrected for effects of topography and past climatic changes.

Based on all this information two regions of special interest were selected for drilling (Fig. 1): one in the Reuss river valley where a Bouguer-anomaly of 5 mgal and seismic surveys indicated the presence of a 250 m deep and 500 m wide, Quaternary-filled canyon and another one near the thermal spring of Bad-Schinznach. At the first location two 70 m deep holes were drilled, one at each flank of the canyon. In these drillholes temperature, flowmeter, gamma-ray, gamma-gamma, neutron-neutron and caliper logs were run. Pumping tests were also performed. The results

indicate that the thermal water in this zone drains from deeper aquifers
into the quaternary sediments at greater depth, probably near the bottom
of the canyon.

At Bad-Schinznach, a network of 20 shallow (5-25 m deep) holes was
established for periodic temperature and electrical conductivity measurements as well as for chemical analyses. These observations enabled the
distinction of several water types. One of them turned out to be strongly
influenced by uprising thermal water. A 135 m deep exploration hole was
located in this zone. The temperature distribution (Fig. 3), the geological description of the cuttings and geophysical log analysis (the
same logs were run as in the Reuss valley holes) provided information on
the porosity/permeability distribution in the series of this drillhole.
Two pumping tests were performed between 70 m and the bottom of the
drillhole. Temperature and mineralization of the water as well as the
yield of the section 72 - 86 m led to commercial utilization of the
pumped thermal mineral water by the spa of Bad-Schinznach.

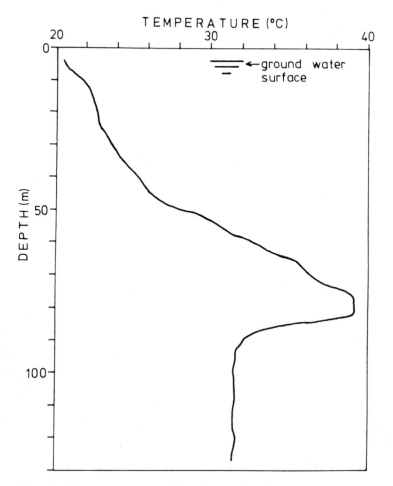

Fig. 3. Temperature profile in Bad-Schinznach. The temperature profile
indicates movements of thermal water along fractured zones of limited
extent.

Geochemical interpretation, geothermal conditions

In order to distinguish the different water types, their electrical conductivity was measured. In the investigated area conductivity values of surface water are below 500 μS/cm at 20 °C (1 μS/cm corresponds to 100 mho/m), whereas water of deep origin, mixed or unmixed with surface water, can have conductivities up to 5400 μS/cm at 20 °C (e.g. water of Baden with up to 4.5 g/l TDS). Water with > 1000 μS/cm at 20 °C and temperatures > 9 °C (mean annual air temperature in the studied area) were further investigated by periodic chemical analyses.

Several water types were identified on the basis of chemical composition (Neff 1980, Vuataz 1980): surface water, water related to the Muschelkalk (which can have very different composition and/or concentration) and water originating from deep circulation through the crystalline basement. Some of the water originating from the crystalline basement is strongly mineralized, due to the percolation through Middle Triassic evaporites during its ascent.

From the chemical analyses, SiO_2-geothermometers for chalcedony and quartz, as well as NaKCa- and NaKCaMg-geothermometers, were calculated (Fournier 1981). Reservoir temperatures up to about 75 °C were obtained (Tab. 1). It has been noted that the SiO_2-geothermometer is strongly influenced by dilution or deep mixing effects (for instance in Bad-Schinznach).

Tab. 1. Reservoir temperatures obtained from the geothermometers (°C).

Thermal area	maximum emergence temperature	T SiO_2 (chalcedony)	T SiO_2 (quartz)	T NaKCa	T Na/K
Baden (springs)	48	70	102	75*	–
Bad-Schinznach (drillhole)	36	37	70	70	–
Zurzach (drillholes)	42	42	75	127	70
Springs from the Muschelkalk	9 – 15	10 – 15	15 – 50	20 – 40	–

* Corrected for Mg

The concentrations of the isotopes tritium, deuterium, oxygen-18 were measured in the most interesting water samples. The interpretation of the results indicates that practically all the thermal waters are more or less diluted by fresh and young waters originating from near surface aquifers. The mean altitude of the recharge area, obtained by the oxygen-18 method, indicates an origin from the north, in the direction of the Black Forest, and not from the Alps, as frequently postulated in the past. This origin appears to be the same for the thermal water of Zurzach and those discharging along the "Hauptüberschiebung".

Regional heat flow (of conductive and convective origin) was mapped using only temperature logs without visible water circulation effects and thus assuming that convection takes place, if at all, below the measured section. Heat flow was determined and analysed for each stratigraphic unit in the drillholes by taking into account more than 200 thermal conductivity measurements, covering the geological units in the investigated area. The regional heat flow distribution indicates values > 120 mW/m^2 in the area between the Black Forest massif to the north and the "Hauptüberschiebung" to the south; further to the south the heat flow decreases to values < 80 mW/m^2.

Mathematical modelling

Two-dimensional models were calculated on two scales:
   a) A regional thermohydraulic model, based on the finite element technique, simulating heat and mass transfer within a cross section running between two water dividers, from the Alps (in the south) to the Black Forest massif (in the north). The results are in agreement with existing information on the regional temperature field and will be published in Neff (1980).
   b) Local hydrothermic models, based on the finite difference technique, simulating the ascent of thermal water along a permeable vertical fault. A section of the 2-dimensional hydrothermic model is given in Fig. 4 (upper part). The calculations lead to quantitative relations between: 1. the temperature of the issuing thermal water at the surface, 2. the spatial extension of the thermal anomaly, 3. its minimum age, 4. the maximum circulation depth of the water and 5. the amount of the water flow (yield per time unit). Two different model curves are represented in Fig. 4 (lower part), showing the calculated temperature distribution at 100 m depth in the vicinity of the fault (x=0) for two depths (Z$_O$) of the aquifer. The temperature value at x=0 represents approximately the spring-temperature at the earth's surface. The initial reservoir is assumed to be located between 1 and 1.5 km depth (assuming no temperature decay due to mixing effects). The shape of the calculated temperature anomaly is not in contradiction with field observations. The amount of circulating water necessary to produce anomalies of this extent is relatively low (about 400 m^3 per meter length of the "Hauptüberschiebung" and per year).

Conclusions

The interpretation of the field observations, supplemented by the modelling results, revealed that concentrated thermal groundwater flow in the discharge area is strongly related to intersections of the "Haupt-überschiebung" with a system of subvertical north - south striking faults, both representing permeable elements (see Fig. 2). Deep groundwater flow (probably in combination with convective cells) between infiltration and discharge area of the thermal water system is supposed to be bound to the subvertical fault system. Seismic and geothermal evidence point towards sufficiently permeable faults or fractures into the crystalline basement, which in the Rheingraben system are known to exist down to a depth of 6 kilometers (Werner & Parini 1980). The amount of water involved in this circulation system and discharging at the surface is relatively limited, however during its upward flow into the discharge area, the deep thermal water is mixing with relatively colder water in the known aquifers, probably at the top of the crystalline basement and in the "Muschelkalk". The resulting total discharge along the "Hauptüberschiebung" within the area under study (approximate length: 20 km) was estimated to be on the order of 250 1/sec on the basis of interpretation and modelling results. The geothermal potential of the thermal water system explored has to be

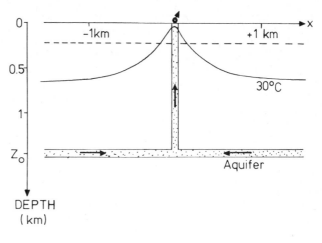

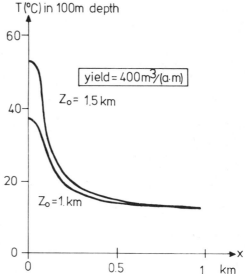

Fig. 4. Hydrothermic model (top) and its result (bottom) for the hot springs along the "Hauptüberschiebung". For details see text.

defined in more detail. However, besides the possibility of a more intensive balneological utilization of the mineralized thermal water, it is supposed that the available amount of thermal water might also be sufficient for limited selected applications, especially in conjunction with the spas, such as space heating in the Schinznach area.

For the methodology of a follow-up geothermal exploration we emphasize here the following aspects: a) more detailed analysis of the mixing in the thermal water system by geochemical techniques and by more refined modelling should be performed, b) more information is expected from additional boreholes in the discharge areas, c) finally, deeper exploration holes at sufficient distance from the discharge area could be recommended, with the possibility of a posterior use of the pumped water.

Acknowledgements. The project was financially supported by the Swiss National Foundation for Energy Research (Nationaler Energieforschungs-Fonds, NEFF) and carried out by the following institutions:
 - Swiss Federal Institute of Technology, Zurich: Institute of Geophysics
 - Swiss Federal Institute of Technology, Lausanne: Institute of Energy Economics and Management (IENER)
 - University of Geneva, Department of Mineralogy
 - University of Neuchatel, Center of Hydrogeology
 - Motor Columbus Consulting Engineers Inc., Baden: Geological Service.
The collaboration of these partners, together with the Thermalquelle AG Zurzach and the Bad-Schinznach AG, allowed an optimal utilization of the existing know-how.

References

Fournier, R.O., 1981: Application of water geochemistry to geothermal exploration and reservoir engineering. - In: L. Rybach & L.J.P. Muffler (Eds.), Geothermal systems - principles and case histories. - Wiley & Sons Ltd., 109-143.

Jaffe, F., Rybach, L. & Vuataz, F.D., 1980: Exploration for low enthalpy geothermal energy in Switzerland. - Proceed. of the UNITAR Conference on long term energy resources, Montreal, Canada. (in press).

Motor Columbus, 1977: Geothermische Studie der Warmwasserzone Zurzach - Lostorf - Baden. - Schriftenr. Eidg. Komm. Gesamtenergiekonzeption, Studie Nr. 14.

Nationaler Energie-Forschungs-Fonds (NEFF), 1980: Geothermische Prospektion im Raume Koblenz - Wildegg - Dielsdorf. (in press).

Schindler, C., 1977: Zur Geologie von Baden und seiner Umgebung. - Beitr. Geol. Schweiz, kl. Mitt. Nr. 67.

Vuataz, F.D., 1980: Seasonal behaviour of the thermal waters from Switzerland and some neighbouring regions. - Proceed. of the 3rd Internat. Symp. on Water - Rock Interaction, Edmonton, Canada, 129-131.

Werner, D. & Parini, M., 1980: The geothermal anomaly of Landau/Pfalz: An attempt of interpretation. - J. Geoph. 48: 28-33.

Geothermal investigation in the deep sedimentary basin of The Netherlands

W. A. Visser

Visser, W.A., 1982: Geothermal investigation in the deep sedimentary basin of The Netherlands. - Geothermics and geothermal energy, eds. V. Čermák & R. Haenel, E. Schweizerbart'sche Verlagsbuchhandlung, Stuttgart: 249-256.

Abstract: Due to about 1400 oil and gas exploration and production wells and an extensive net of seismic lines subsurface geology is reasonably well known to depths of 3 to 4 km. As a result of depositional and diagenetic circumstances lower Permian and lower Triassic sandstones are the main targets, with possibilities in top lower Carboniferous and lower Cretaceous.
As in the larger part of the country gradients of 3 $^{\circ}$C/100 m are the rule, the development of geothermal resources depends on the presence of suitable reservoir rocks at appropriate depths. The inventory and mapping of aquifers is underway, not only at greater depths, but also, for the purpose of seasonal storage and recovery of energy, at depths between 400 and 1000 m.
Temperatures from oil wells have been corrected and compiled by S. Prins (1980); his maps are published in the Subsurface Temperature Atlas of Europe. The results of temperature measurements in groundwater observation wells (50-400 m) and of gravimetric and magnetic surveys in the horst and graben area in the SE will be reported on resp. by W.J. van Dalfsen and J.W. Bredewout, both in this book.
A demonstration project, intended to produce 100 $^{\circ}$C saline water from a Middle Bunter sandstone (2900 m) at Spijkenisse (SW of Rotterdam), is planned. A feasibility study and a detailed seismic survey have been completed. Drilling will be decided on at a later stage. A production/reinjection doublet is envisaged, possibly in combination with a shallow (700 m) storage/recovery doublet.

Author's address: Geological Advisor, Netherlands Organisation for Applied Scientific Research TNO, P.O. Box 285, 2600 AG Delft, The Netherlands

Introduction

In The Netherlands the interest in the possibilities of low-caloric geothermal energy dates from 1974, when a group of specialists on geosciences and geotechniques was convened by the vice-president of the Netherlands Organization for Applied Scientific Research TNO. The purpose of this group, the Discussion Group Geothermal Energy, was to evaluate the feasibility of exploitation of geothermal energy in The Netherlands. In 1976 the Group reached the conclusion that a thorough evaluation of geothermal propects was highly desirable, and consequently a program of investigation was compiled (1977).
    In the meantime the European Communities (EC) had opened the possibility to obtain financial aid on projects on energy research and development and

on projects on waste heat conservation (period 1976-1979). As a follow-up
in 1979 financial aid could be applied for so-called demonstration
projects, i.e. on drilling and exploitation of geothermal energy, and on
conservation. Moreover, a second program on research and development was
initiated (period 1980-1983). Here the Discussion Group found opportunity
to fulfil a coordinating and stimulating task, and on all EC-program
tenders for subsidy on various projects were submitted. The financial aid
offered by the European Communities may amount to 30 to 50 % of actual
costs; the remainder has to be found elsewhere. In 1979 the Minister of
Economic Affairs allotted funds for investigation and research, if a
National Program Geothermal Energy were submitted. This National Program
1979 comprises those items of the 1977 program of investigation that have
high priority to reach a well-founded evaluation of prospects. It consists
of five parts (Visser 1979):
1. A demonstration project,
2. Evaluation of the potential,
3. Underground storage and recovery of energy,
4. Evaluation of geothermal prospects,
5. National information centre geothermics.

Geological review

In a deep sedimentary basin, as present in The Netherlands, where low-
caloric geothermal energy only is to be expected, investigation is aimed
at the mapping of permeable formations and of the temperature fields in
the subsurface. The Netherlands form part of the extensive European basin,
in which a thick series of mainly marine sediments has been accumulated.
Subsidence did not proceed at a uniform rate and periods of local and
regional emergence were followed by renewed subsidence (Heybroek 1974).
This resulted in an extremely complicated geological subsurface structure,
as is expressed in age, thickness, extent and facies of the still present
sediments and in the gaps in their succession. In general diagenetic
changes in sediments and formation waters cause deterioration of reservoir
properties in the former and greatly enhanced salinities of the latter.
It is certain that in many places circumstances will be unfavourable for
the exploitation of geothermal energy, while elsewhere the production of
highly saline waters will necessitate special measures to be taken.
    Geothermal gradients vary. They are low (1.4 to 2 °C/100 m) in shallow
Quaternary and Pliocene deposits due to ground water flow. In the deeper
subsurface they show a connection with lithology (i.e. thermal conductivity), varying between 1.8 and over 5 °C/100 m in respectively the Permian
rocksalt formation and the Carboniferous (Visser 1978); in the larger
part of the country gradients are "normal", 3 °C/100 m.
    Due to exploration and exploitation of petroleum and natural gas, the
subsurface geology is reasonably well-known to depths of 3 to 4 km, i.e.
from Lower Carboniferous onward. Some 1400 deep boreholes and a detailed
net of seismic lines have been carried out. To complete the geothermal
program collaboration of the oil industry is indispensable; fortunately
this collaboration has been obtained through the State Geological Survey.

Production

Due to the rises of the price of crude oil, in the near future, from a
non-economic or at best marginal commodity, low-caloric geothermal energy

may become competitive. It is of importance that the source is situated
in or near the consumer's area, for instance in a newly built town
quarter designed for block heating or near an agricultural area.

The formation water contained in the pores of a permeable sedimentary
layer is the only medium by which geothermal energy can be carried to the
surface. Because of the high salinity of these waters, disposal at the
surface after cooling should be avoided. In France this problem has been
solved by reinjection of the cooled waters into the producing formation
in a closed circuit containing heat exchangers. It entails the drilling
of two holes, a so-called doublet, and involves the spending of electricity
to operate the pumps. From oil-production practice it appears, as a rule,
that injectivity index is about a factor three higher than the productivity
index, i.e. the amount of fluid that can be injected into, respectively
produced from, the reservoir from, respectively into, a well per unit
time and per unit pressure difference between the reservoir and the bottom
of the well. This may affect the energy balance adversily. Further, anti-
corrosion measures are almost certainly necessary.

Additional advantages of reinjection are maintenance of pressures and
of the material balance in the subsurface. Production of geothermal energy
causes cooling of the formation at a far greater rate than can be re-
plenished by the natural heat flow. Therefore, geothermal energy is an
exhaustible mineral resource with a restricted duration. Estimates vary
from some 20 to over 50 years. The rock particles will have a considerable
contribution to the ultimate amount of energy produced; on its way from
reinjection to production well the "water front" will precede the "cold
front" (Sauty et al. 1980). However, inhomogeneities in the reservoir
formation may cause unpleasant surprises.

The demonstration project

In 1960 in the course of the exploration for oil and gas, a boring was
put down to a depth of over 3000 m in the municipality of Spijkenisse SW
of Rotterdam. It encountered at a depth around 2900 m an about 120 m thick
sandstone of the Bunter Formation of lower Triassic age, forming part of
a WNW-ESE trending zone of extremely high porosity (up to 24 %). A
restricted (extending over only 8 m of the bed) drill-stem test produced
saline water (100 g/l) of 100 °C. This test was not conclusive as to the
final capacity of this sandstone; in the most optimistic estimate it may
be as high as 200 m³/h; it may be half of that amount or even less.
Anyhow, it was considered an indication sufficiently encouraging to
proceed, and to apply for subsidy (1973).

In Spijkenisse an extensive building program is under way, in which
the application of geothermal energy for heating and household purposes
could very well be realized. On the other hand the use of waste heat from
the nearby industrial complex along the waterway to Rotterdam is also
being envisaged. The possibility of combination of these two sources of
energy is being studied.

The first phase of the project consists of a feasibility study with
the admittedly scant data available, an analysis of costs and profits
concerning both, money and energy, and the design of the surface installation. In the study consideration has been given to the comparison with
other forms of energy and to ways of seasonal storage and recovery,
including possibilities in a shallow Eocene sandstone known to be present
in the borehole of 1960 mentioned above.

The second phase is a detailed seismic survey. Since it had to be

carried out in and around a built-up and highly populated area, the vibro-
seis method was employed. Although the prospective sandstone was encountered in a borehole, its extent and structure (the area is known to
be faulted) are insufficiently known from the older wide-grid seismic
surveys. The field work was carried out in February 1980; processing is
at present still under way. No results are yet available.

Further consideration of reservoir properties, the results of the
feasibility study and of the seismic survey may indicate, whether minimum
requirements for a viable prospect are present. Much depends on the hydrological properties of the reservoir sandstone. They are insufficiently
known from the 1960 well and can only be properly investigated by drilling.
This, the third phase of the project, the execution of which will be
decided upon at a later stage, consists of a borehole to a depth of 3050 m.
Deviation to 1300 m from the vertical will be necessary. The site of the
old borehole and its surrounding prospective area are covered by a
recently completed quarter of the town; moreover, striking the prospective
sandstone at the upthrown side of a major fault, known to be present, but
not its exact location, must be avoided. In this borehole the usual
investigations will be made, and extensive production tests will be
carried out, if successful followed by reinjection tests. The produced
water will be chemically analysed.

Only after conclusion of above investigation and interpretation of the
results, the economic viability of the project can be evaluated properly.
If positive, the fourth phase will be executed, viz. the drilling and
testing of a reinjection well and installation of pumps and surface
equipment.

Evaluation of the potential

In the National Program Geothermal Energy the following investigations
and studies are included:
   1. subsurface temperatures,
   2. inventory of reservoir formations,
   3. geophysical surveys.

1. Subsurface temperatures

Since 1976 temperatures in shallow (up to 400 m deep) ground-water
observation wells have been measured systematically by the Groundwater
Survey TNO in Delft. A subsidy of the Commission of the European
Communities (CEC) was obtained. As a result the shallow subsurface
temperature fields are well-known; a series of temperature maps at
intervals of 25 m have been compiled at depths between 25 and 250 m. The
measurements and their results are discussed by van Dalfsen (1977).

Groundwater Survey TNO will also do research to determine in-situ
thermal conductivities in unconsolidated sediments in boreholes. A CEC-
subsidy has been granted.

In about 400 deep hydrocarbon-exploration wells temperatures were
measured before thermal equilibrium was established; in several oil and
gas fields reliable temperature data exist. The former need correction.
A first effort was made by the Working Group for Petroleum Technique,
University of Technology, Delft; later a general correction method was
applied by the State Geological Survey. This resulted in the compilation

of more or less reliable subsurface temperature maps between 500 and 3000
m depths with intervals of 500 m. Temperatures at these depths, where the
influence of the hydrological cycle has disappeared, are largely
determined by the different thermal conductivities of the sediments
present. No indication of enhanced heat flow from greater depths have
been found (Prins 1980).

2. Inventory of reservoir formations

For the evaluation of the feasibility to apply low-caloric geothermal
energy, knowledge of reservoir formations is of paramount importance, viz.
mapping of their lithological and hydrological properties, their extent,
thickness, localities and depths. Data from deep boreholes and from
seismic sections are being compiled on subsurface maps. The work is
carried out under auspices of the State Geological Survey by a retired
petroleum geologist, G. Milius; assistance on geothermal and geohydrologi-
cal aspects is rendered by the Groundwater Survey TNO. A CEC-subsidy is
granted.

Due to the known subsurface temperatures reservoirs are considered at
depths greater than 2000 m, or, since temperatures between 50 and 60 °C
may be acceptable also, greater than 1000 m. Reservoirs consist of sandstone of Lower Cretaceous, Lower Triassic and Lower Permian ages, while in
the Dutch-Belgian border area a cavernous limestone at the top of the
Lower Carboniferous forms another prospect.

3. Geophysical surveys

In the horst and graben area in the SE part of the country (Peel area) a
discrepancy between gravimetric and magnetic anomalies exists that from
known geological structure could not be explained. Moreover, in the
southern Peel area and in a nearby mine in the Federal Republic of
Germany abnormal high degrees of coalification occur, while indications
of rather high geothermal gradients are locally present. A detailed
gravimetrical and magnetical survey was commenced in 1977 by the Geophysical Department of the Vening Meinesz Laboratory, State University in
Utrecht. A CEC-subsidy was granted in 1980. The measurements and their
results are discussed in this book by J.W. Bredewout.

In northern Belgium, on the flank of the Brabant Massif a stratigraphic
hiatus exists between Namurian and Lower Carboniferous Visean limestones
(Bless et al. 1976). In the top of the latter a level of cavernous lime-
stones is present with good reservoir properties that in the Dutch-
Belgian border area is situated at a depth of about 2250 m (Gulinck 1956).
Temperature is rather high, about 100 °C (Legrand 1975).

In the course of this year further investigation is carried out by the
Belgian Geological Survey by drilling and a seismic survey, for which a
CEC-subsidy is granted. It is hoped that, due to the differences in
seismic characteristics of dense and cavernous limestone, the latter can
be mapped. A supplementary survey on Dutch territory, closing a gap
caused by the course of the international border, has been decided on.
For this work also, to be carried out simultaneously and in close cooperation with our Belgian colleagues, a CEC-subsidy is available.

If the seismic method proves successful in distinguishing the two
kinds of limestone, further application in structurally higher areas,
where the Visean is expected to be present within reasonable depth limits,
will be envisaged.

Underground storage and recovery of energy

The importance of seasonal storage of energy in the form of warm water
involves two aspects, viz. conservation of energy and decrease of thermal
pollution. Overproduction of a geothermal doublet during summer or super-
fluous warm water from industrial and electricity plants could be stored
underground and recovered during winter (Visser 1976).

Apart from economic viability, which has still to be ascertained,
precautions, due to the geological and hydrological characteristics of
the subsurface, are to be taken. The stored waters should not contaminate
any fresh ground water suitable for human consumption and agriculture.
This means storage below a poorly pervious, thick and extensive clayey
bed in an environment of saline formation water; in The Netherlands in
Tertiary sands, which, at depths from 400-500 to about 1000 m, are present
over large areas. To create the necessary space injection is to be carried
out under pressure. The pressure should not exceed a certain value, because
in that case the overlying protective clay bed may be fractured. This
restriction, together with porosity and permeability, extent and thickness, determine storage capacity. The relative mobility of stored and
formation water determine the shape of the injected water body. The less
dense and less viscous injected waters tend to float upward and to follow
streaks of better permeability. For the conservation of energy, it is of
importance, however, that the injected water remains in a body as compact
as possible replacing formation water to a great degree. Deterioration of
the reservoir bed and its hydrological characteristics will almost certainly result, if chemical and physical equilibrium is disturbed by injected
water of a composition different from that naturally present in the
aquifer. For above reasons it is envisaged that through heat exchangers
at the surface the temperature of the formation water is increased by the
energy to be stored. Here also, a doublet of boreholes that establish a
closed circuit of formation water appears attractive.

The investigation will be carried out jointly by State Geological
Survey and Groundwater Survey TNO; the former will attend to geological,
sedimentological and geophysical aspects, the latter to geohydrology and
geothermics. The investigation comprises:
- inventory and mapping of Tertiary sands between 400 and 1000 m depth
  from borehole and seismic data; their extent and thickness and reservoir properties; quality and extent of covering poorly pervious
  beds; composition of formation waters; subsurface temperatures,
- study of hydrological aspects, such as maximal allowable injection
  pressures, injection rates and capacities of the reservoir sands;
  shape and extent of the injected water bodies,
- recovery of the stored energy, comprising study on the percentage
  recoverable and temperature after one and more cycles of storage and
  recovery; technique of storage and recovery,
- feasibility of production of moderately warm Tertiary formation waters
  and temperature increase through heat pumps.

In the course of the investigations, which will take several years,
prospects for (a) pilot project(s) will certainly be found. One of these
may be in conjunction with the Spijkenisse demonstration project. Execution will provide the answer to feasibility and to technical and economic
merits.

Apart from storage and recovery the investigations have a great
scientific interest. In this interval of the geological column the
transition occurs from fresh or brackish ground water, which belongs to
the present hydrological cycle, to saline and virtually stagnant formation

waters, separated from the cycle. It is a "no man's land", too deep to be of interest to the drinking-water companies, too shallow to the oil industry.

Evaluation of geothermal prospects

From the results of the above surveys, expected to be completed by 1983, areas will be selected that merit further investigation of the possibility to produce geothermal energy and of seasonal storage underground. The experience obtained in the Spijkenisse demonstration project will be of great value for the proper execution of such projects. Since, however, in these selected areas geological circumstances are expected to be different from those at Spijkenisse, a similar procedure should be applied, i.e. before drilling a feasibility study and a detailed seismic survey, should be carried out.

National information centre geothermics

By the Groundwater Survey TNO in Delft an information centre will be founded. In the past years books and articles have been collected. A further extension of this library and an abstract-attending service are envisaged to be established in the course of 1980.

Conclusion

It is to be expected that geothermal energy and underground storage and recovery will constitute a not negligible supplement to energy require-ments and saving of conventional forms of energy. At present the potential of geothermal energy is difficult to estimate; preliminary estimates, however, indicate a few per cents of present-day requirements comparable to the possible contribution of solar or wind energy. The saving by underground storage may be in the same range or even better. The shallow aquifers possess greater extent, thickness and more favourable character-istics than the deep ones; on the other hand underground storage is still in the research stage.

References

Bless, M.J.M., Bouckaert, J., Bouzet, Ph., Conil, R., Cornet, P., Fairon-Demaret, M., Groessens, E., Longerstaey, P.J., Meessen, J.P.M.Th., Paproth, E., Pirlet, H., Streel, M., van Amerom, H.W.J. & Wolf, M., 1976: Dinantian rocks in the subsurface North of the Brabant and the Ardenno-Rhenish Massifs in Belgium, The Netherlands and the Federal Republic of Germany. - Med. Rijks Geol. Dienst, 27, No.3: 81-195.
Dalfson, W. van, 1977: Temperature measurements in shallow observation wells in the Netherlands. - Seminar on Geothermal Energy, Brussels 6-7-8 Dec. 1977, Direct. - Gen. for Research, Science and Education, EUR 5920, Vol. 1, 99-110.
Gulinck, M., 1956: Caractéristiques hydrogéologiques du sondage de Turnhout. - Comm. Sém. Phys. Globe, sc. 12 Oct. 1956: 1-6.
Heybroek, P., 1974: Explanation to tectonic maps of The Netherlands. - Geol. Mijnb., 53 (2): 43-50.

Legrand, P., 1975: Jalons géothermiques. - Verh. Belg. Geol. Dienst, no. 16: 1-46.

Prins, S., 1980: In: Atlas of subsurface temperatures in the European Community. Compiled by R. Haenel, Commission of the European Communities.

Sauty, J.P., Gringarten, A.C., Landel, P.A. & Menjoz, A., 1980: Life time optimization of low enthalphy doublets. - Sec. Internat. Sem. Results EC Geoth. Energy Res., Strasbourg, 4, 5, 6, March 1980: 273-278.

Visser, W.A., 1976: Hydrogeological considerations on underground storage of low-caloric energy. - Mem. Internat. Hydr. Ass. XI, Budapest Conference: 793-800.

— , 1978: Early subsurface temperature measurements in The Netherlands. - Geol. Mijnb. 57 (1): 1-10.

— , 1979: De mogelijkheden van aardwarmte in Nederland. - De Ingenieur, Jg.91, Nr.46: 804-810.

Geothermal resources assessment

E. Gosk

with 2 figures and 1 table

Gosk, E., 1982: Geothermal resources assessment. - Geothermics and geothermal energy, eds. V. Čermák & R. Haenel, E. Schweizerbart'-sche Verlagsbuchhandlung, Stuttgart: 257-263.

Abstract: The preliminary evaluation of the main geothermal para-meters leads to the conclusion that many of the Danish geothermal reservoirs can compete with the traditional energy sources only if a rather optimistic estimate is made about the transmissivity. Because of the high salinity of geothermal water and the necessity of maintaining reservoir pressure, the geothermal doublet is considered. The value of geothermal resources depends mainly on a combination of: reservoir temperature, reservoir transmissivity and the necessary drilling depth. Each of the above mentioned parameters can be a limiting factor for use of the reservoir for geothermal energy production. An alternative approach, where these three main parameters are combined into one dimensionless parameter, is proposed to define geothermal resources and reserves. A few examples are given to illustrate the proposed method.

Author's address: Geological Survey of Denmark, Thoravej 31, DK-2400 Copenhagen NV, Denmark

Introduction

It is known that, in general, temperature increases with depth and the typical rate of increase (geothermal gradient) is around 30 °C/km. Consequently, an enormous amount of heat is stored in the rock masses. Simple calculations show, that $8 \cdot 10^{17}$ joule ($\triangleq$ 19 mill. tons of oil) is stored for every km² if the heat stored down to 5 km is considered (geo-thermal gradient = 30 °C/km, heat capacity of the rocks = $2.1 \cdot 10^{15}$ joule/ (km³ °C), and the reference temperature for the calculations is surface temperature).

If the reference temperature is chosen to be 100 °C, the heat content corresponds to 3 mill. tons of oil.

In spite of the huge amount of energy existing in the subsurface, two main disadvantages make geothermal energy less attractive: 1 - An "energy carrier" which can transport sufficient amounts of heat to the surface is needed (that is normally steam or/and water present in a certain kind of porous or fissured rock formation called a reservoir). 2 - The temper-ature of the energy carrying fluid is in many cases not sufficiently high for electric power generation which limits the use of geothermal energy. Furthermore, the geothermal fluids contain rather small amounts of energy per mass unit (0.5 to 5 % when compared with oil).

Classification and definitions of different forms of geothermal energy and methods of assessment of geothermal resources and reserves have been provided by several authors (P. Muffler & R. Cataldi 1977, P. Muffler 1978, P. Muffler & R. Christiansen 1978).

In this paper, a method of assessment of geothermal resources of the type which is typical for Denmark (hot water reservoirs in sedimentary basins, normal geothermal gradient), is presented. It is shown that three main parameters characterizing geothermal reservoirs (transmissivity, temperature and depth) may be combined into one dimensionless number-power factor - which may be used to describe the investigated reservoir and/or to compare different reservoirs. The Danish reservoirs are expected to have transmissivities ranging from poor to intermediate and the geothermal water is highly saline. In that situation a geothermal doublet has to be used to prevent environmental pollution and to avoid pressure depletion and resulting decrease in the productivity.

Reservoir dynamics

Water movement in the geothermal reservoir is induced by pumping from the production well and reinjecting the cooled water into the injection well of the geothermal doublet. For a given flow rate, the power consumption for pumping and reinjection will depend mainly on the transmissivity of the reservoir, $K \cdot h$, and can be expressed as:

$$P_{pump} = 2 \cdot Q \cdot Q / (2 \pi K \cdot h) \; \ln(D/r_w) \cdot \rho_w \cdot g / \eta \tag{1}$$

where:

P_{pump} = power required for pumping and reinjection, watt (thermal)
Q = pumping rate = reinjection rate, m^3/s
K = hydraulic conductivity of the reservoir, m/s
h = thickness of the reservoir, m
D = distance between doublet wells, m
r_w = radius of the pumping/reinjection well, m
ρ_w = water density, kg/m^3
g = acceleration due to gravity = $9.81 \; m/s^2$
η = factor used for conversion of electrical power into thermal power (= 0.4).

For practical purposes, the value of $\ln(D/r_w)$ may be approximated by 9, and ρ_w may be set equal to 1000.
Eq. (1) then simplifies to:

$$P_{pump} = 7 \cdot 10^4 \cdot Q^2 / (K \cdot h) \tag{2}$$

Eq. (2) is a good approximation for homogenous, isotropic reservoirs but for geothermal doublets it should be used only for comparison of different reservoirs, rather than for exact calculations. The transmissivity value, $K \cdot h$, is strongly dependent on temperature and may vary by a factor of 4 for hot and cooled water. Since the region occupied by cooled water will increase with time, the power required for pumping and reinjection will also increase. By using the transmissivity value corresponding to cooled water, a conservative estimate of the reservoir performance can be made. Eq. (2) could be rather easily extended to take into account frictional losses in the casing, heat exchanges, pipes, etc., but that would require additional assumptions. That extention however would not be justified because of the simplifications previously made.

Main parameters for reservoir assessment

There are three main parameters describing the geothermal reservoir:
- transmissivity of the reservoir
- temperature
- depth

Transmissivity (hydraulic conductivity x thickness) describes the ability of the reservoir to produce geothermal water and determines the power requirement for pumping and reinjection.

Temperature of the geothermal water determines the thermal output of the doublet.

Depth of the reservoir is essential when drilling costs are taken into consideration.

Each of the above mentioned parameters can be a limiting factor in the use of geothermal reservoirs for energy production. In order to make comparisons among different geothermal reservoirs easier, the three main parameters will be combined into one single dimensionless "power factor" f:

$$f = \text{POWER OUTPUT/POWER INPUT} \tag{3}$$

Power output, P_{out}, can be defined by a simple equation, (4), while power input is a sum of two terms: power required for pumping and reinjection, P_{pump} - defined by Eq. (2), and doublet cost expressed in units comparable with P_{out} and P_{pump} - watt thermal.

$$P_{out} = \rho_w \cdot C_w \cdot Q \cdot \Delta T = 4.2 \cdot 10^6 \cdot Q \cdot \Delta T \tag{4}$$

where: $\rho_w \cdot C_w$ = heat capacity of water = $4.2 \cdot 10^6$ joule/(^oC m³)

ΔT = useable temperature difference

P_{out}, Q as defined before.

The remaining quantity - doublet cost expressed as power consumption - requires further explanation. Normally, a geothermal doublet is designed for a period of 20-30 years (in the following, 25 years is assumed to be the life time). The method of calculating its life time under various reservoir conditions is described in several papers (Gringarten & Sauty 1975, Gringarten 1978). Doublet cost consists mainly of drilling - and maintenance cost. It is known that drilling costs increase in a way similar to that for energy prices, and expressing drilling costs in terms of tons of oil equivalent may then be a reasonable assumption. That amount of energy divided by the life time of the doublet gives "power consumption equivalent to drilling costs". A similar approach is used for expressing maintenance costs in power units. Power input is then equal to the sum of power required for pumping and reinjecting, power consumption equivalent to drilling costs and power consumption equivalent to maintenance costs.

Power balance for geothermal doublet

To illustrate the use of the power factor defined by Eq.(3), several examples are calculated and the results are shown in Fig. 1 and Fig. 2.

Three hypothetical geothermal doublets are considered and for each of them three different transmissivities (corresponding to poor, intermediate and good reservoir) are assumed. Tab. 1 describes these nine cases together with the maximum value of the power factor, f_{max} for each case.

Fig. 1 gives a graphical representation of Eq.(2) - Pumping power and

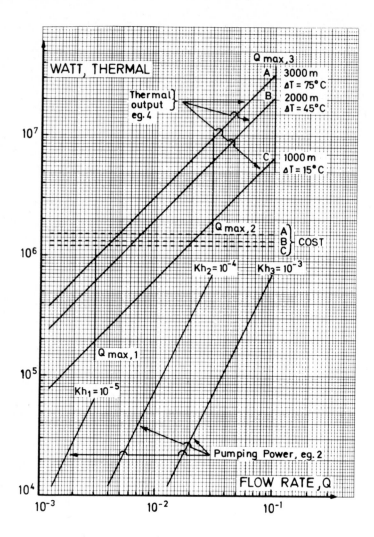

Fig. 1. Pumping power (pumping and reinjection) and thermal output of a
geothermal doublet as a function of flow rate, Q. A, B and C correspond
to different depths (1, 2 and 3 km) and different temperatures of the
reservoirs (100, 70 and 40 °C). Reinjection temperature is assumed to be
25 °C and the useable temperature difference, ΔT, is then 75, 45 and
15 °C for A, B and C respectively. Transmissivities $K \cdot h_1$, $K \cdot h_2$, $K \cdot h_3$
determine maximum flow rates $Q_{max,1}$, $Q_{max,2}$ and $Q_{max,3}$. Drilling and
maintenance costs are assumed to be independent of flow rate.

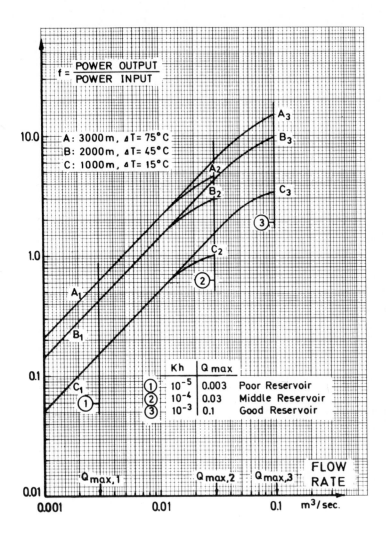

Fig. 2. Power factor, f, as a function of flow rate, Q. Indices 1, 2 and 3 correspond to different transmissivities. For moderate flow rates ($Q \leq 0.015$ m³/s) f-value is independent of reservoir transmissivity, kh.

Eq.(4) – Thermal output as a function of the flow rate (pumping/reinjection rate). Maximum flow rate for the poor reservoir ($K \cdot h_1 = 10^{-5}$ m²/s) is determined by the maximum drawdown (chosen to be 400 m), equal to $3 \cdot 10^{-3}$ m³/s (= 11 m³/h). The maximum drawdown for the intermediate reservoir is $Q_{max,2} = 3 \cdot 10^{-2}$ m³/s (= 110 m³/h) while for the good reservoir ($K \cdot h_3 = 10^{-3}$ m²/s), the maximum flow rate $Q_{max,3} = 10^{-1}$ m³/sec (= 360 m³/h) is determined by the performance of the well and surface equipment, rather than by the maximum drawdown.

   Fig. 2 gives a graphical representation of Eq.(3) and combines the input from Tab. 1 and results from Fig. 1. A constant geothermal gradient of 30 °C/km is assumed, reinjection temperature is 25 °C and surface temperature is 10 °C. When heat losses are neglected, useable temperature differences, ΔT, will be 75 °C, 45 °C and 15 °C for doublet A, B and C respectively. When reinjection temperatures are changed and/or heat losses are included, the value of the power factor, f, will be changed proportionally with the new useable temperature difference.

Tab. 1. Input data for the calculation of f-factor (Eq.(3)).
Assumptions:
 - reinjection temperature = 25 °C
 - life time of a doublet = 25 years
 - cost of oil = $ 220/ton (= $ 35/barrel)
 - heat capacity of oil = $4.2 \cdot 10^{10}$ joule/ton
Index 1, 2 & 3 corresponds to poor, intermediate and good reservoir respectively.

Doublet	Depth	Temp.	Drilling cost	Maintenance cost	Doublet cost	Transmissivity	f-max.
	m	°C	$ 10^6	$ 10^6/year	MWatt	m^2/s	
A_1	3000	100	4	0.1	1.5	.00001	0.6
A_2	3000	100	4	0.1	1.5	.0001	4.4
A_3						.001	14.3
B_1						.00001	0.4
B_2	2000	70	3	0.1	1.3	.0001	2.9
B_3						.001	9.5
C_1						.00001	0.2
C_2	1000	40	2.5	0.1	1.2	.0001	1.0
C_3						.001	3.3

Conclusions

Estimation of the parameters of geothermal reservoirs is rather difficult. If the temperature and the depth of the reservoir can be determined with reasonable accuracy, the estimation of transmissivity may be incorrect by a factor of 10 or even more. Hence simple methods should be used to make a preliminary assessment of geothermal resources and reserves.
    From the examples shown (Fig. 2), it may be concluded that the geothermal doublet, producing from reservoirs where transmissivity is around $10^{-5} m^2$/s (1 Darcy-meter, cold water), can hardly be regarded as feasible because the f-value is far below one. The intermediate reservoir, with $kh_2 = 10^{-4} m^2$/s, requires the useable temperature difference to be higher than 15 °C in order to give f-values higher than one. In case B_2, where $f_{max} = 2.9$, a change in reinjection temperature from 25 °C to 45 °C and heat losses corresponding to 5 °C will reduce f_{max} from 2.9 to 1.3.

Acknowledgements. The author wants to express his gratitude to his colleagues from the Advisory Commitee on Project Management, Geothermal Energy and particularly to Dr. I.P. Ungemach and to Dr. R. Haenel for critical remarks and fruitful discussion.

References

Gringarten, A.C., 1978: Reservoir lifetime and Heat Recovery Factor in
    Geothermal Aquifers used for Urban Heating. - Pageoph, 117: 297-308.
Gringarten, A.C. & Sauty, I.P., 1975: A theoretical study of heat
    extraction from aquifers with uniform regional flow. - J. Geophys. Res.
    80(35).
Muffler, P. (ed.), 1978: Assessment of Geothermal Resources of the
    United States. - Geological Survey Circular 790.
Muffler, P. & Cataldi, R., 1977: Methods for Regional Assessment of Geo-
    thermal Resources. - Paper presented at Lardarello Workshop on Geo-
    thermal Resource Assessment and Reservoir Engineering.
Muffler, P. & Christiansen, R., 1978: Geothermal Resource Assessment of
    the United States. - Pageoph, 117: 160-171.

Subsurface temperature distribution in Western Czechoslovakia and its mapping for appraising the exploitable sources of geothermal energy

Vladimír Čermák and Jan Šafanda

with 2 figures and 1 table

Čermák, V. & Šafanda, J., 1982: Subsurface temperature distribution in Western Czechoslovakia and its mapping for appraising the exploitable sources of geothermal energy. - Geothermics and geothermal energy, eds. V. Čermák & R. Haenel, E. Schweizerbart'sche Verlagsbuchhandlung, Stuttgart: 265-270.

Abstract: All available data on the temperature loggings were used to propose a pattern of the subsurface temperature distribution. The maps of the isobaths of temperatures 130 °C and 180 °C were constructed for the western part of Czechoslovakia. These maps are briefly discussed in terms of the prediction of prospective areas of geothermal potential.

Authors' address: Geophysical Institute, Czechoslovak Academy of Sciences, 141 31 Praha, Czechoslovakia

Introduction

The world-wide crisis in energy prices was a strong stimulus for research and development of alternative energy sources, among them also for the study of the possibility to use the terrestrial heat, i.e. the geothermal energy. One of the most important criteria of any prospective geothermal area is the increased value of the terrestrial heat flow which reflects high subsurface temperature at accessible depths. In the preliminary assessment of the regional economic potential of the economically usable geothermal resources in a given area, the maps showing the underground temperature distribution are necessary. For the industrial use of the geothermal energy usually the temperature of 150 °C is often mentioned, with the lower limit of 100 °C. During exploiting and transporting of the heated medium from depth to the surface certain cooling must be expected. For the 30 °C cooling down the underground temperature values of 130 and 180 °C were chosen and for these values the corresponding depths were calculated. The regional maps of the isobaths of the temperatures 130 and 180 °C were constructed for the territory of western Czechoslovakia, i.e. for the Bohemian Massif and for the adjacent Carpathian Foredeep on its eastern rim.

Temperature data

According to the available information the temperature of 130 °C was reached by direct drilling on the above mentioned territory only in a single case in the Jarošov-1 hole at depth of approximately 5200 metres. However, the temperature logging in this hole was taken only one day after the drilling had been ceased and the temperature field in the hole was thus not in equilibrium and the heat flow value was not calculated. In all other holes the drilling reached shallower depths and the depths of the temperatures of 130 and 180 °C are to be extrapolated. All data were summarized by Čermák et al. (1979) together with the measured temperature at maximum depth reached by the probe, extrapolated temperatures, mean thermal conductivity, mean temperature gradient, heat flow value, co-ordinates, etc. for each hole.

Temperature in the boreholes is usually measured by logging at the last measurement after several days of total equilibrium. The minimum length of the recovery time before the temperature log is given by the instructions of the Czechoslovak Geological Survey (Ústřední ústav geologický, 1959). The aim of these instructions is, however, to ensure a good technical temperature record with the accuracy of ± 0.5 °C, which not necessarily must reach higher standards required by the heat flow determination. Therefore for heat flow data calculation certain selection was applied on temperature records (Čermák 1978) or the equilibrium temperature profile was re-calculated using the repeated temperature logging technique described by Lachenbruch & Brewer (1959).

Maps of isobaths of temperature 130 °C and 180 °C

For a known value of the surface heat flow (Q) and the coefficient of the thermal conductivity (k) the temperature at given depth (z) can be determined:

$$T(z) = T_o + Q \int_0^z \frac{dz}{k} \, ,$$

where T_o is the surface temperature. For all individual points at which heat flow value was determined, the corresponding depths of temperatures 130 and 180 °C: h_{130} and h_{180} were calculated using the local values of heat flow and thermal conductivities of rocks collected from the particu-lar hole. As the rocks drilled through in different holes may vary considerably and the measured value of the thermal conductivity in a relatively shallow hole characterizes more or less the local properties of a cover rather than the basement rocks, the calculated isobaths h_{130} and h_{180} may show great scatter with little relation to the regional field.

On the basis of numerous determinations of the thermal conductivities of different rocks (Čermák 1967) and according to their statistical occurrence in view of the stratigraphical and lithological structure of the individual geological formations, the characteristic thermal conducti-vities were adapted (Tab. 1) and used for the extrapolation of measured temperature profiles. Evaluating the mean thicknesses of the individual layers the revised depths of the temperatures 130 and 180 °C were determined: H_{130} and H_{180}. These latter values were used for the con-struction of the maps of the isobaths (Fig. 1 and 2). Each individual value was used as the check point and the course of the isolines then adapted to the local tectonic structure and the existing heat flow map (Čermák 1978).

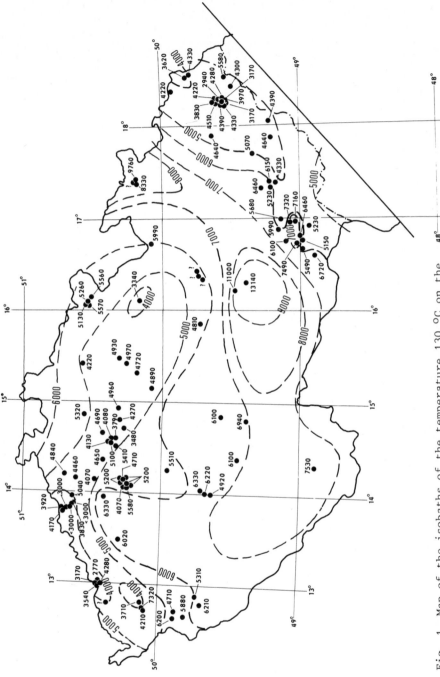

Fig. 1. Map of the isobaths of the temperature 130 °C on the territory of western Czechoslovakia.

Tab. 1. Characteristic thermal conductivities of rock samples from selected geological and/or tectonic units on the territory of western Czechoslovakia.

Geological or tectonic unit	Number of measurements	Thermal conductivity $(Wm^{-1}K^{-1})$
Cretaceous Basin:		
Mesozoic sediments	136	1.85
Permocarboniferous sediments	162	2.35
Coal basins in the Bohemian Massif:		
Kladno-Rakovník Basin	110	2.28
Žacléř-Svatoňovice Basin	39	2.64
Ostrava-Karviná Basin	328	2.78
Krušné Hory Mts. region:		
Teplice porphyry	22	2.5
Cínovec granite, greisen	32	2.7
Jeseníky Mts. region:	74	2.82
Bohemian Massif - intramontane stable block:		
crystalline schists, gneiss	146	2.67
granite	83	2.72
Carpathian Foredeep (southern and central parts):	236	2.66
Vienna Basin (Neogene)	36	1.53

Conclusions

The minimum depths of the temperature 130 °C are in the range of 2500-3000 metres and correspond well to the regions of high heat flow (more than 100 mWm^{-2}) in the area of Teplice in northern Bohemia. However, high heat flow values observed here may be caused by local underground water movement and the surface heat flow values may thus not be reliable for depth below 1000 metres. Therefore it would be better to limit ourselves to the isobath of 4000 m as characterizing the "warmest" areas, hopefully prospective for future exploration of the geothermal energy. These warmest areas are located in both tectonically disturbed zones of the Bohemian Massif, i.e. along the Krušné Hory Mts. and in its graben and along the axis of the Cretaceous Basin in northeastern Bohemia. The broader region of future interest should thus include the whole Krušné Hory rift zone, the Cretaceous Basin and its northern margins and probably also the eastern slopes of the Bohemian Massif in the area of the Ostrava-Karviná Coal Basin, i.e. all the regions bounded by the isobath of 5000 metres in the Fig. 1.

In the prevailing part of the Bohemian Massif and in the Carpathian Foredeep the temperature 130 °C is at depths of 5000-6000 metres and below. In the "coldest", so called stable intramontane block of the massif, which is of the Pre-Variscan origin, the temperature of 130 °C can be found at the depth of 7000 metres. The similar general pattern is typical also for the isobathic map of temperature 180 °C, in which the coldest part of the Bohemian Massif is characterized by isobaths of 9000 - 11000 metres (Fig.2).

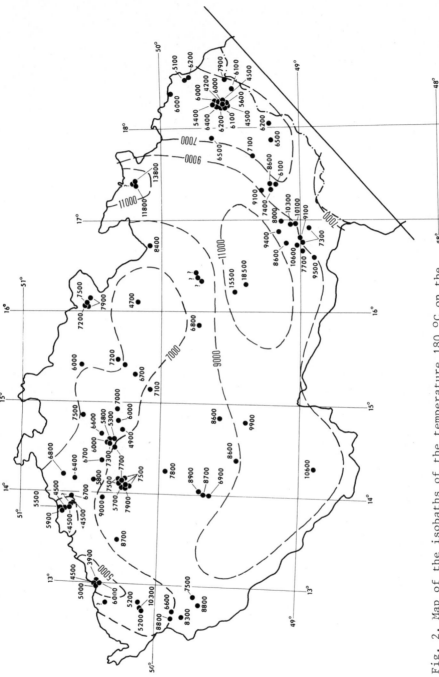

Fig. 2. Map of the isobaths of the temperature 180 °C on the territory of western Czechoslovakia.

From the point of view of utilizing the geothermal energy the only prospective areas can be looked for in the Krušné Hory Mts. zone (the Ohře rift) and in the western part of the Cretaceous Basin. Anyhow, even here the favourable temperatures for mass production of energy are relatively very deep and any future use of the geothermal energy in Czechoslovakia is thus a question of the economic calculations including the drilling and hydrofracturing costs. More prospective is the use of the geothermal energy for space heating and in the agriculture, rather than for production of electricity.

The accessible resource base in northern and northeastern Bohemia can be estimated to be 28×10^{21} joule to the depth of 6 km and 95×10^{21} joule to a depth of 10 km. The steady heat supply by the terrestrial heat flow is 2.1 GW. The current project of the geothermal exploration is aimed to produce an underground heat exchanger in dry rocks by hydrofracturing till the end of 1983.

References

Čermák, V., 1967: Coefficient of thermal conductivity of some sediments and its dependence on density and water content of rocks. – Chemie der Erde 26: 271-278.

– , 1978: First heat flow map of Czechoslovakia. – Trav. Inst. Géophys. Tchécosl. Acad. Sci., No 461, Geofysikální sborník 1976. Académia, Praha: 245-261.

Čermák, V., Šafanda, J. & Halada, S., 1979: Hloubky teplot pro energetické využití v ČSR. – In: T. Pačes (Ed.), Možnosti využití tepla suchých hornin v ČSR. – Ústřední ústav geologický, Praha: 16-22. (Czech.)

Lachenbruch, A.H. & Brewer, H.C., 1959: Dissipation of the temperature effect of drilling a well in Arctic Alaska. – Geol. Surv. Bull. N-1083 C, 73 pp.

Ústřední ústav geologický, 1959: Předepsané prostoje vrtu před provedenou teplotní karotáži na strukturních vrtech. – Geofond, Ústřední ústav geologický, Praha, 15 pp. (Czech.)

Utilization of geothermal low enthalphy water in the southern part of the Federal Republic of Germany

J. Fritz, R. Haenel and J. Werner

Fritz, J., Haenel, R. & Werner, J., 1982: Utilization of geothermal low enthalphy water in the southern part of the Federal Republic of Germany. - Geothermics and geothermal energy, eds. V. Čermák & R. Haenel, E. Schweizerbart'sche Verlagsbuchhandlung, Stuttgart: 271-273.

Abstract: The "Geothermal Demonstration Project Saulgau" is thought to be the first step in investigating the possibilities for utilizing the enthalphy below the pre-Alpine Tertiary Basin. Hydro-geological and geothermal data for the Saulgau region: karst aquifer in Upper Malm, 350 m thick, covered by 550 m Tertiary sediments, low salinity, well-head temperature 40 °C, temperature gradient down to 650 m:4.8 °C/100 m, expected production rate 50-100 l/s. The extractable heating power amounts to 13 MW or 24 GWh/a; with the use of heat pumps: 33 GWh/a. The heat will be used for heating purposes.

Authors' addresses: J. Fritz, Ing.-Büro Fritz, Am Schönblick 1, 7417 Urach, Federal Republic of Germany; R. Haenel, Niedersächsisches Landesamt für Bodenforschung, Stilleweg 2, 3000 Hannover, Federal Republic of Germany; J. Werner, Geologisches Landesamt Baden-Württemberg, Albertstraße 5, 7800 Freiburg, Federal Republic of Germany

The Project

The "Geothermal Demonstration Project Saulgau" was approved at the end of 1979.

The project Saulgau is designed to serve as a pilot- and demonstration project to show the efficiency of this resource for a combined supply for heating purposes, drinking water and industrial water for mixed consumer structures.

The present regional hydrogeological exploration, as well as the technical preparations for the planned deep boreholes are in progress.

Geoscientific base

Geology

Saulgau is situated near the northwestern margin of the South German Molasse Basin. The dip of the pre-Tertiary formations is directed from the Swabian Jura toward the Alps; that is, from northwest to southwest. Consequently, the Upper Malm that forms the stratum step of the Swabian Jura is covered in the Saulgau area by Tertiary Molasse sediments more than 500 metres thick.

Geothermics

Saulgau is situated within a small positive geothermal anomaly, which has probably no connection with the well-known Urach anomaly. Its origin is not yet clear.

The temperature gradient of the undisturbed temperature field down to 650 m depth, is about 4.8 °C/100 m and the well-head temperature about 40 °C.

Hydrogeology

The pre-Tertiary strata in the southwest German section of the Molasse basin contain two efficient aquifers: the granular limestone of the Upper Malm with a thickness of up to 350 metres, and the Upper Muschelkalk which is about 800 metres deeper with a thickness of about 65 metres. Both aquifers are completely separate ground water systems with their own circulation.

Geoscientific problems and measures to solve them

Definition of the problems

In the case of water with low enthalphy, a thermal station would quickly produce unfavourable results if noticeable cooling took place within a short time, because the technical concept is based on the primary data. For that reason the questions concerning the constancy of temperature and volume of available groundwater, together with their modifications as a function of time, are of great importance.

Thermal water borehole Saulgau 1

The first evidence from the Saulgau field resulted from the thermal water borehole Saulgau 1 with a final depth of 650 m in the karstic limestones of the Upper Malm in 1977. During the pumping test (maximum throughout of pumps: 576 m³/h) no reduction was measurable. The maximum water temperature is 41.9 °C.

Productivity and efficiency

The technical efficiency will be tested by further pumping tests in the existing deep borehole of Saulgau 1 and by pumping tests in the planned deep boreholes of Saulgau 2 and Saulgau 3. The project is based on a high output of 100 l/s. The tests will be completed by measurements to determine the aquifer parameters, namely transmissivity and storage co-efficient.

Possibilities of utilization

Quantity of geothermal energy

On the presumed base of a production rate of 360 m³/h and a temperature difference of 32 °C the heating power amounts to 13 MW. 1800 hours of maximum heating consumption are assumed, the result is a geothermal heat output of 24 GWh/a.

Energy for heating

We intend to install a central heating plant using compressor heat pumps to obtain a higher level of temperature. Altogether five of heat pump units are planned. These systems will work more economically if they are designed with the evaporators in series instead of in parallel. The thermal energy output will be 33 GWh/a.

Economical aspects

Cost of project

The following breakdown of cost (prices in German Mark for January 1979) does not include the cost of research.

1. Geothermal borehole Saulgau 2	DM 3 430 000.--
2. Central station	DM 5 870 000.--
3. Groundwater recharge	DM 1 640 000.--
4. Transmission pipelines	DM 3 260 000.--
total project cost	DM 14 200 000.--
total operating cost per year	DM 1 433 000.--

Cost of energy

In order to break even, the management has to calculate 4.4 Deutsche Pfennige per kilowatthour, on the basis of an annual supply of 33 giga-watthours of energy for heating. Thus, if full use were made of the geo-thermal energy produced at Saulgau, the plant would work economically.

Summary

The "Geothermal Demonstration Project Saulgau" is thought to be the first step in investigating the possibilities for utilizing the enthalphy below the pre-Alpine Tertiary Basin.

The following questions are to be answered: Are there hydraulic connections? What is the karst-water flow pattern? How can the exchange between rock and water be quantified? How much water and heat can be extracted without lowering the temperature? What is the smallest possible distance between productive wells? The hydrogeological and geothermal research, combined with the project data obtained, serve to furnish proof that the supposed usable geothermal energy in the Upper Swabian Molasse Basin amounts to about 24 terawatthours.

Deep investigation of the geothermal anomaly of Urach

H. G. Dietrich, R. Haenel, G. Neth, K. Schädel and G. Zoth

Dietrich, H.G., Haenel, R., Neth, G., Schädel, K. & Zoth, G., 1982:
Deep investigation of the geothermal anomaly of Urach. - Geo-
thermics and geothermal energy, eds. V. Čermák & R. Haenel, E.
Schweizerbart'sche Verlagsbuchhandlung, Stuttgart: 275.

Authors' addresses: H.G. Dietrich, Stadtwerke Urach, Rathaus,
7432 Urach, Federal Republic of Germany; R. Haenel and G. Zoth,
Niedersächsisches Landesamt für Bodenforschung, Stilleweg 2,
3000 Hannover, Federal Republic of Germany; G. Neth and K. Schädel,
Geologisches Landesamt Baden-Württemberg, Albertstraße 5,
7800 Freiburg, Federal Republic of Germany

Until recently the exploration of the geothermal anomaly of Urach was
limited to water-bearing sedimentary horizons, mostly in the Upper
Muschelkalk of the Middle Triassic. For deep investigation the Urach 3
borehole was drilled. Originally this borehole was planned to a depth of
2100 m or 2500 m, but the final depth is 3334 m. The geological profile
is as follows (depth in brackets): Quaternary (13 m), Dogger (282 m),
Lias (401 m), Keuper (665 m), Muschelkalk (843 m), Buntsandstein (908 m),
Rotliegend (1486 m), Carboniferous (1604 m), and the basement (3334 m)
which can be subdivided into three units: orthogneiss, paragneiss and
mica-syenites, or metablastite, paragneiss and diatexite respectively.
   All water-prospective horizons were tested by pumping or drill stem
tests, in both cased and open hole. The Upper Muschelkalk yields more than
12.6 l/s (pump capacity) with a specific yield of about 3.5 l/s·m, the
Muschelkalk/Buntsandstein transition zone and a fissured zone at 1775 m
yield maximum 1.3 l/sec and 3.3 l/s. All other horizons produced less than
0.1 l/s.
   The measured temperatures were 68 °C at 1000 m, 103 °C at 2000 m,
134 °C at 3000 m, and 143 °C at the final depth. Irregularities in
temperature are observed from the surface to about 1000 m depth; they are
probably caused by water movement behind the casing of the borehole.
Additional ascending water from great depth is indicated by the measured
temperature graph.
   After a short phase of interpretation of the technical and geological
parameters, the suitability of the basement rocks for the hot dry rock
concept was tested. We succeeded in fracturing the basement rocks in four
sections. Although only one single well is available at Urach, it was
possible to produce small amounts of geothermal energy by circulating
water within the tubing-fracture-annular-system.

The full text is published in:

Dietrich, H.G., Haenel, R., Neth, G., Schädel, K. & Zoth, H., 1980:
   Deep investigation of the geothermal anomaly of Urach. - Proc. of
   2. Internat. Sem. on Results of EC Geothermal Energy Research, Strasbourg, 4-6 March, 1980, 253-265.
Haenel, R., 1980: Geophysikalische Untersuchungen in der Forschungs-
   bohrung Urach. - Report, NLfB-Hannover, Archiv No. 85 805: 1-75.

Seismic-refraction studies of the geothermal area of Urach, Southwest Germany

Claus Prodehl, Dieter Emter and Martin Jentsch

with 2 figures

Prodehl, C., Emter, D. & Jentsch, M., 1982: Seismic-refraction studies of the geothermal area of Urach, Southwest Germany. - Geothermics and geothermal energy, eds. V. Čermák & R. Haenel, E. Schweizerbart'sche Verlagsbuchhandlung, Stuttgart: 277-283.

Abstract: In 1978-79 seismic-refraction experiments were carried out in order to investigate the crustal structure of the geo-thermally anomalous area of Urach, Southwest Germany. This paper summarizes the results which relate to the depth structure and the velocity distribution of the Variscan basement as derived from first arrivals of compressional waves. Application of the MOZAIC time-term analysis resulted in a detailed map of depth isolines of the basement which outline the extent of the SW-NE striking Rot-liegend trough with maximum depths of 1.6 - 2.0 km around Urach. A uniform velocity distribution within the basement has been de-rived with the velocity placed at 5.66 ± 0.02 km/s. No significant decrease in velocity within the anomaly as compared to that outside could be detected. No evidence for velocity anisotropy was found in the area. It is suggested that any high temperature source for the anomaly has to be located at depths well below 3 km.

Authors' addresses: C. Prodehl, Geophysikalisches Institut der Universität Karlsruhe, 7500 Karlsruhe, Federal Republic of Germany; D. Emter, Observatorium Schiltach, 7620 Wolfach, Federal Republic of Germany; M. Jentsch, BEB, 3000 Hannover, Federal Republic of Germany

The Urach geothermal area is located at the northern edge of the Swabian Alb southeast of Stuttgart in Southwest Germany. It reveals an oval form with its axis running in WSW-ENE direction, as is illustrated by the lines of equal geothermal steps after Carlé (1974) in Fig. 1.

As part of a multidisciplinary approach, in July, 1978 and May, 1979 seismic-refraction surveys were carried out making use of specially designed borehole shots as well as commercial quarry blasts. Thus 10 fully or partly reversed profiles were observed (Fig. 1), some of which have been recorded at a distance up to 150 km. The surveys and data preparation are described in detail by Jentsch (1980) and Jentsch et al. (1980, 1981).

The principal objective of the seismic-refraction investigations was to define the topography and the P-wave velocity structure of the crystalline basement. Studies in areas of higher temperatures have re-vealed clearly distinguishable effects on seismic velocities (Majer & McEvilly 1979). However, for a low-grade thermal anomaly such as that in the Urach area such changes will be rather small, as has been shown by laboratory measurements (e.g., Hughes & Maurette 1956, Timur 1977), and therefore much more difficult to detect. Nevertheless, the methods used in this study are thought to provide a single test for such effects by

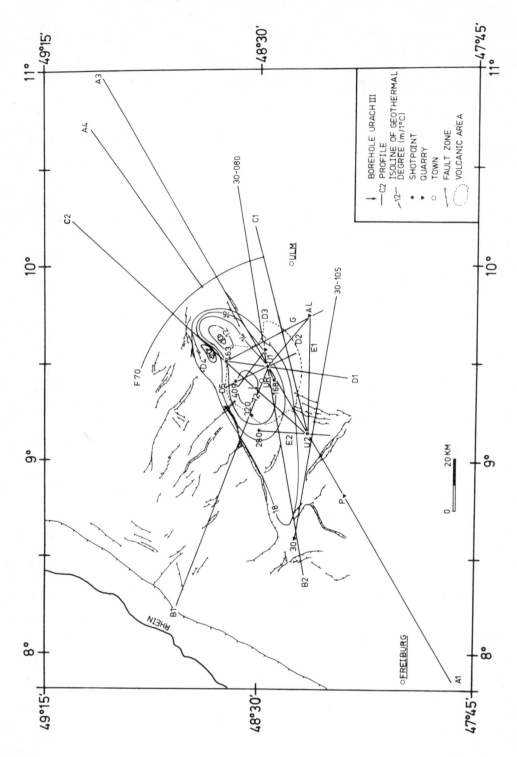

allowing for variations in the refractor velocity (gradient, anisotropy etc.).

The whole area is covered by Mesozoic sediments. The cliff of the White Jurassic (Malm) Plateau which runs parallel to the axis of the geothermal anomaly divides the area into two parts: the Swabian Alb proper in the south with an elevation of 700 - 850 m NN (above mean sea level) consists of a 250 - 400 m thick layer of White Jurassic limestones covering softer sediments of Jurassic and Triassic age (Dogger, Lias and Keuper) which are exposed in the foreland to the north at a surface elevation of only 300 - 450 m NN (Schädel 1977). About 300 Upper Miocene volcanic chimneys of the Swabian Volcano (Geyer & Gwinner 1968, Mäussnest 1974), which pierce the Mesozoic sediments, have been located in the area.

The basement consists of granites and gneisses of Variscan age (Schädel 1977). As has been suggested by earlier geological and geophysical investigations (Breyer 1956, Emter & Prodehl 1977, Schädel 1977), the basement forms the so-called Schramberger trough. This trough is filled by Upper Permian (Rotliegendes) sediments which are absent under most parts of the Swabian Alb.

To achieve the objective of deriving the basement structure, only first arrivals of waves refracted in the crystalline basement were used. These were analyzed by means of Hagedoorn's (1959) plus-minus method to obtain a first idea of the refractor velocity. The velocities thus obtained show no systematic tendency toward significantly higher or lower values within the anomaly or outside.

The whole data set was subsequently treated by the MOZAIC time-term analysis as developed by Bamford (1976). This method allows a joint inter-pretation of all data in the presence of a laterally heterogeneous structure and permits the consideration of velocity variations within the refractor such as gradients or anisotropy (Bamford 1976, Bamford et al. 1979). The topography of the basement has been derived. In the pursuit of this analysis, furthermore, the refractor has been modeled to have either uniform velocity or a small vertical velocity gradient. As a result, on a statistical basis, a uniform velocity of 5.66 ± 0.02 km/s is the best solution for the crystalline basement in the Urach area. A lower boundary of this 5.66-km/s layer can be placed at a depth of 4 - 5 km, as a rough estimate on the basis of unreversed profiles where phases originating from depths below the basement were observed.

The experience with data sets of much weaker, i.e. uneven, azimuthal distribution of observations than the one at hand proves that rather weak anisotropy (2 - 3 %) even at great depths ($\sim$ 40 km) can be detected (Bamford et al. 1979). Therefore the present data set should permit a reliable test for the presence of anisotropy. As a result, solutions allowing for the presence of velocity anisotropy within the basement were less satisfactory than those that did not. Thus it can be stated that, within the limits of measurement error ($\sim$ 1 %), the basement in the Urach area is isotropic.

Fig. 1. Schematic map of the 1978-79 seismic-refraction surveys. Isolines of equal geothermal steps (geothermal degree in m/°C) after Carlé (1974).
Profiles 1971-77: 30-080, 30-105
profiles 1978-79: A1, A3, A4, B1, B2, C1, C2, D1 - D5, E1, E2, G
fan with radius 70 km from shotpoint U2: F70
quarries: Al Allmendingen, P Plettenberg, Z Zainingen, 30 Sulz
shotpoints 1978-79: U1, U2, 168, 280, 320, 409, 463
town of Urach: UR

The presence of lateral velocity variations was tested by calculating
a velocity for each observation point and plotting it at the midpoint
between shotpoint and corresponding recording station. An attempt to
identify systematic changes of velocity throughout the area of investigation inside and outside the geothermal anomaly proper by contouring lines
of equal velocity was unsuccessful (Jentsch et al. 1980, 1981).

To convert the delay times τ obtained in the MOZAIC analysis to depths
H to the basement the simple relation

$$H = \frac{V \cdot \overline{V}}{(V^2 - \overline{V}^2)^{1/2}} \cdot \tau$$

can be used, where V is the refractor (basement)

velocity and $\overline{V}$ a mean overburden velocity. Such a computation gives a
depth beneath the surface for each observation point. A summarizing contour
map using a mean velocity of 3.6 km/s for the overburden and showing
depths of the basement beneath the surface has been published by Jentsch
et al. (1980). However, if strong variations of the topography are present,
as in the area of investigation, it is preferable to use a constant
reference depth (e.g. 500 m above sea level). Therefore, the delay times
shown by Jentsch et al. (1980, 1981) have been corrected to a constant
station elevation of +500 m NN (above mean sea level). For the correction
a velocity of 3.9 km/s, which represents an average velocity for the
Jurassic formations between station and reference level was used. The
corrected delay times were then converted to depths with the use of the
above formula. The mean velocity of 3.6 km/s used for the overburden
between reference level and basement is based on velocity values given for
the sediments by Breyer (1956) and downhole velocity measurements in the
Urach 3 research borehole (Wohlenberg 1981). It should be noted that the
delay times computed for points at the edge of the survey are prone to
systematic errors and therefore have not been used.

In Fig. 2 the resulting depths are plotted in the form of isolines
graded in 200 m intervals. If allowance is made for the 0.02 - 0.05 s
uncertainty of the delay times, which corresponds to a depth uncertainty
of 100 - 200 metres, there seems to be reasonable agreement with the
available borehole data (Geyer & Gwinner 1968, Schädel 1978, Wohlenberg
1981) and earlier measurements (Breyer 1956).

The basic result of the present investigation is the contour map of
the crystalline basement (Fig. 2). As overall feature, this map provides
an outline of extension and borders of the Rotliegend trough mentioned
above. The greatest depths of 1.6 - 1.8 km below +500 m NN are obtained
southeast of Urach. It should be noted that maximum basement depths do not
coincide with the maxima of the geothermal anomaly.

The results lead to the following conclusions: If any velocity changes
are present in the basement of Urach and if they are due to thermal effects,
then, on the basis of laboratory measurements (e.g. Hughes & Maurette
1956), they are below the resolving power of our method and cannot exceed
1 - 2 %. This agrees with the temperature measurements in the Urach 3
borehole (Wohlenberg 1981). The temperature increase of less than 50 °C
within the basement of the borehole, as compared to normal, corresponds to
less than 1 % change in P-velocity (Meissner et al. 1980). If the origin
of the geothermal anomaly is a high-temperature source, it is certainly
located below 4 - 5 km in depth. Otherwise, it would exert stronger effects
on the seismic velocities. Any crack or fissure systems that can serve as
aquifers for geothermal convection within the basement are likely to have
random distributions. Regular crack patterns should have had a bulk
anisotropic effect on the seismic velocities (Bamford & Nunn 1979).

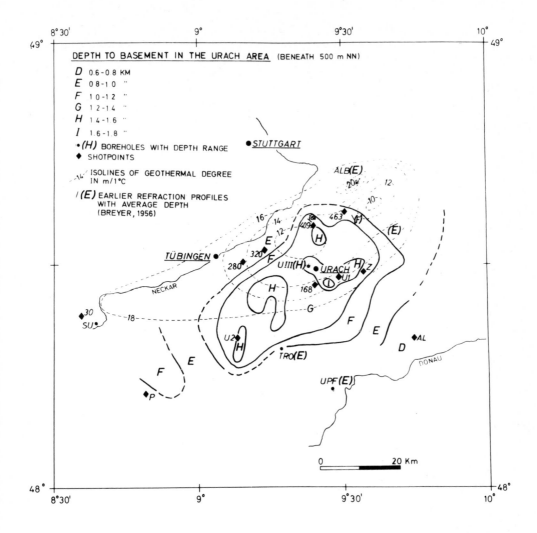

Fig. 2. Contour map of basement depths below +500 m NN graded in 200 m intervals (heavy lines). Isolines of equal geothermal steps after Carlé (1974).
Explanations of shotpoints see Fig. 1
boreholes: ALB Albershausen, SU Sulz, TRO Trochtelfingen, U III Urach 3, UPF Upflamör.

With the type of experiment and the methods applied, a straightforward and comparatively cheap way deriving refractor velocity structure and topography as related to compressional waves over a large area is given. This provides basic information to aid in determining where other, more detailed methods, e.g., reflection seismic studies (Meissner et al. 1980) should be applied and other geophysical methods such as electric or magnetic investigations are likely to add significant information.

Acknowledgements. The investigation was funded by a joint project of the Commission of the European Communities (EG) under contract no. 071-76 EG G/A-23-5 and the German Federal Ministry for Research and Technology (BMFT) under contract no. ET 4027 A, as part of a multidisciplinary project headed by Dr. R. Haenel.
  Dr. D. Bamford, now at British Petroleum, London, took particular interest in the project during field work and interpretation. We grate-fully acknowledge the aid of Dr. K. Fuchs. W. Kaminski supplied continuous improvements of the data processing system at ghe Geophysical Institute. Model computations were carried out at the Computer Centre of the Univer-sity of Karlsruhe. We thank all participants in the field experiments for their personal engagement and effort which contributed so much to the success of the experiments. The following institutions participated: Universities of Bochum, Frankfurt, Göttingen, Karlsruhe, Kiel, Stuttgart, ETH Zürich, Niedersächsisches Landesamt für Bodenforschung and Geologi-sches Landesamt Nordrhein-Westfalen. For the organisation of the shots the help of the Gewerkschaften Brigitta und Elwerath Betriebsführungsgesell-schaft m.b.H. Hannover, Mr. C. Behnke of Niedersächsisches Landesamt für Bodenforschung, Hannover and of Mr. Kleinmann and Mr. Dudenhöffer of the Landesbergamt Baden-Württemberg at Freiburg is greatly acknowledged. The personal engagement of Dr. Dietrich, Urach, and Dr. Schädel, Freiburg, is highly appreciated. The communities of Burladingen-Melchingen (1978) and Urach (1979) kindly provided the headquarters for the field experiments and much other support.

References

Bamford, D., 1976: MOZAIC time-term analysis. - Geophys. J. R. astr. Soc. 44: 433-466.
Bamford, D., Jentsch, M. & Prodehl, C., 1979: P_n anisotropy studies in northern Britain and the eastern and western United States. - Geophys. J. R. astr. Soc. 57: 397-430.
Bamford, D. & Nunn, K., 1979: In situ seismic measurements of crack aniso-tropy in the carboniferous limestone of northwest England. - Geophys. Prospect. 27: 322-338.
Breyer, F., 1956: Ergebnisse seismischer Messungen auf der Süddeutschen Großscholle besonders im Hinblick auf die Oberfläche des Varisticums. - Z. deutsch. geol. Ges. 108: 21-36.
Carlé, W., 1974: Die Wärme-Anomalie der mittleren Schwäbischen Alb (Baden-Württemberg). - In: J.H. Illies & K. Fuchs (Eds.), Approaches to Taphrogenesis. - E. Schweizerbart'sche Verlagsbuchhandlung, Stutt-gart: 207-212.
Emter, D. & Prodehl, C., 1977: Use of explosion seismics for the study of the Urach area. - Proc. Seminar on geothermal energy, Brussels, December 6-8, 1: 351-366.
Geyer, O.F. & Gwinner, M.P., 1968: Einführung in die Geologie von Baden-Württemberg. - E. Schweizerbart'sche Verlagsbuchhandlung, Stutt-gart: VIII, 228 pp.
Hagedoorn, J.G., 1959: The plus-minus method of interpretation of seismic refraction sections. - Geophys. Prospect. 7: 158-182.
Hughes, D.S. & Maurette, C., 1956: Variation of elastic wave velocities in granites with pressure and temperature. - Geophysics 21: 277-284.
Jentsch, M., 1980: A compilation of data from the 1978-79 Urach, Baden-Württemberg, seismic-refraction experiment. - Open-file report 80-1, Geophysical Institute, University of Karlsruhe, 92 pp.

Jentsch, M., Bamford, D., Emter, D. & Prodehl, C., 1980: Structural study of the Urach area by deep refraction seismics. - Commission of the European Communities, Second International Seminar. Advances in European Geothermal Research, 4-6 March 1980, Strasbourg, 223-230.

− − − − , 1982: A seismic-refraction investigation of the basement structure in the Urach geothermal anomaly, Southern Germany. - In: R. Haenel (Ed.), The Urach Geothermal Project. - E. Schweizerbart'sche Verlagsbuchhandlung, Stuttgart.

Mäussnest, O., 1974: Die Eruptionspunkte des Schwäbischen Vulkans. - Z. Dtsch. Geol. Ges. 125: 23-54, 277-352.

Majer, E.L. & McEvilly, T.V., 1979: Seismological investigations at the Geysers geothermal field. - Geophysics 44: 246-269.

Meissner, R., Bartelsen, H., Krey, Th., Lüschen, E. & Schmoll, I., 1980: Combined reflection and refraction measurements for investigating the geothermal anomaly of Urach. - Commission of the European Communities, Second International Seminar. Advances in European Geothermal Research, 4-6 March 1980, Strasbourg, 231-234.

Schädel, K., 1977: Die Geologie der Wärmeanomalie Neuffen-Urach am Nordrand der Schwäbischen Alb. - Proc. Seminar on geothermal Energy, Brussels, December 6-8, 1977, 1, 53-60.

− , 1978: Bohrung Urach 3, Erdwärmebohrung, vorläufiges Schichtprofil (1:1.000). Geologisches Landesamt Baden-Württemberg, Freiburg.

Timur, A., 1977: Temperature dependence of compressional and shear wave velocities in rocks. - Geophysics 42: 950-956.

Wohlenberg, J., 1982: Seismo-acoustic and geoelectric experiments within the Urach 3 borehole. - In: R. Haenel (Ed.), The Urach Geothermal Project. - E. Schweizerbart'sche Verlagsbuchhandlung, Stuttgart.

Detecting velocity anomalies in the region of the Urach geothermal anomaly by means of new seismic field arrangements

R. Meissner, H. Bartelsen, T. Krey and J. Schmoll

with 8 figures

Meissner, R., Bartelsen, H., Krey, T. & Schmoll, J., 1982:
Detecting velocity anomalies in the region of the Urach geothermal
anomaly by means of new seismic field arrangements. - Geothermics
and geothermal energy, eds. V. Čermák & R. Haenel, E. Schweizerbart'sche Verlagsbuchhandlung, Stuttgart: 285-292.

Abstract: Combined seismic reflection and refraction measurements
with extended spreads have revealed many details of horizontal
(and vertical) interval velocities. Horizontal velocities show a
minimum up to 3 % in the upper 3 km of the crystalline basement
below the Urach geothermal anomaly. Stacking velocities and interval velocities indicate a large low velocity zone of up to 10 %
velocity deviation extending down to the bottom of the crust below
the Urach anomaly. This area also exhibits a relative maximum in
crustal thickness.

Authors' addresses: R. Meissner and H. Bartelsen, Institut für
Geophysik der Universität Kiel, 2300 Kiel, Federal Republic of
Germany; T. Krey and J. Schmoll, PRAKLA-SEISMOS GmbH, 3000 Hannover,
Federal Republic of Germany

Introduction

Laboratory investigation of various materials under high-pressure and
high-temperature conditions have shown that a change of temperature at
constant pressure gives roughly a 2 % change in v_p-velocity for a
temperature change of 100 °C (Kern 1978, Meissner et al. 1930). This value
is valid for p-T conditions in continental crusts. Effects of inhomogeneity
and anisotropy may be much stronger, and careful investigations have to be
carried out, in order to distinguish between these effects. On the other
hand, a velocity variation of 1 % can be measured by special arrangements
of a seismic field survey.
    In order to determine such small lateral velocity variations down to
30 km depth different requirements have to be fulfilled for seismic reflection and refraction work. For reflection work large move-out times are
required, e.g. about 1 s for 10 s two-way travel time. These large move-out
times can only be obtained by continuous spread length of about 23 km.
Multiple coverage is required in order to enhance the signal to noise
ratio. For refraction work also a comparably large spread length and a
high coverage are essential, and for great penetration depth additional
measurements are required at greater distances. A reflection coverage of
- say - 8, as used for the Urach field set-up, results in a refraction
coverage of 16. Fig. 1 shows the field set-up of the two profiles in the
Urach area. Line I consisted of a geophone line of about 23 km length
with three 48 trace independent reflection units. An 8-fold coverage for
reflection and a 16-fold coverage in refraction was achieved by arranging

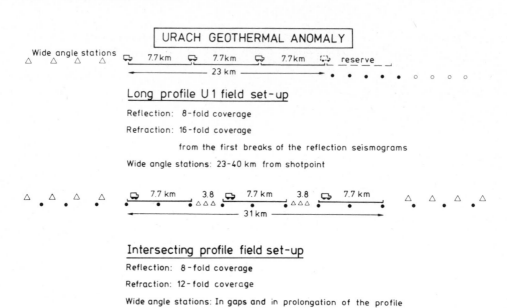

Fig. 1. Field setup of the two intersecting profiles of the Urach area.

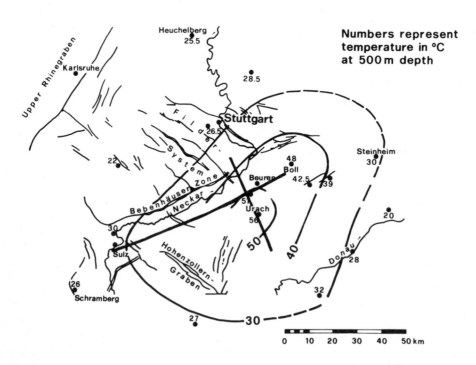

Fig. 2. Tectonic map of SW Germany with location of profiles I and II and isotherms in 500 m depth.

shotpoints at 1.3 km interval. For the picking of the first refracted
arrivals and some wide-angle arrivals the reflection line was extended to
more than 100 km by 27 portable refraction stations. A slightly different
arrangement was made for line II crossing line I nearly perpendicularly.
Fig. 2 shows the positioning of the two lines with regard to the known
heat flow information.

A basic interpretation of the refraction work

The refraction work consists of the evaluation of the first arrivals of
the reflection set-up and the evaluation of arrivals of the 27 refraction
stations. Applying the Wiechert-Herglotz method to corrected and smoothed
travel time diagrams all along the profile I, the isotaches of Fig. 3
could be plotted. As the refracted rays travel approximately horizontally
through the subsurface, Fig. 3 shows the horizontal P-velocities along
the profile. One of the two zones of small velocities coincides with the
Urach area, the other one with the area of Sulz where CO_2 exhalations are
reported. The steep gradient on the right hand side of the profile might
show either a change in material or anisotropy or may be due to insufficient coverage of the profile. From Fig. 3 it is clear that the
velocity anomaly reaches depths of at least 3 km in the crystalline basement with a maximum lateral anomaly at about 1000-2000 m depth. In this
depth range there is a 100 m/s velocity anomaly with regard to the center

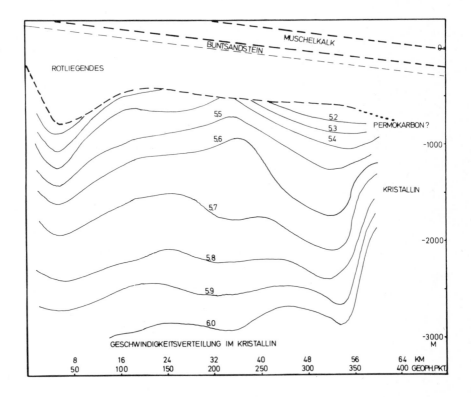

Fig. 3. Isotaches of horizontal velocities in the crystalline basement of
profile I.

of the profile which is the double amount as that calculated before. Averaged over the whole depth range of 3 km, the value of about 50 m/s, corresponding to 50° temperature difference, agrees with the theoretical value. About 7 record sections in the wide angle range have been plotted so far from the refraction stations of the institutes. Fig. 4 gives an example of the data quality between 50 and 110 km distance. Within this range often a double reflection from the lower crust and the crust-mantle boundary appears in the sections. The critical angles are between 55 and 75 kilometers. A small dip towards the ENE is observed on the western part and on the extension of the profile, indicating the rising flanks of the Rhinegraben system. The central data from these refraction studies will be incorporated into the reflection section, shown in the next paragraph.

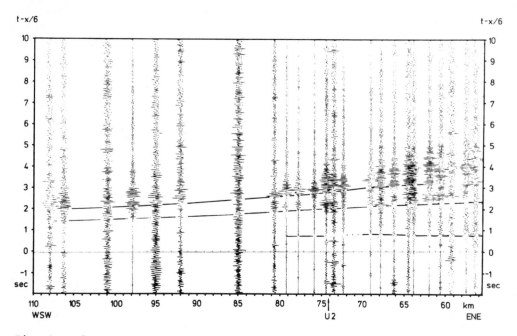

Fig. 4. Seismogram section of a portable station with arrivals between 61 and 110 km.

A basic interpretation of the reflection work

After a very careful and time-consuming correction procedure, involving many checks of the static and dynamic corrections, several stacked record sections were plotted. Fig. 5 shows a section with a 2-fold vertical exaggeration in the average. The strong reflectivity of the lower crust (Meissner 1967) is clearly defined. Undisturbed Moho-reflections over more than 20 km can be found. The dip towards the ENE is very pronounced at the western part of the profile. It changes to the opposite direction just below the center of the Urach anomaly. Here, strong indications of faulting are observed. An updoming in the middle crust is seen at the left part of the profile, probably caused by diapirs. The process of the formation of the Rhinegraben flanks seems to extend into the Urach area,

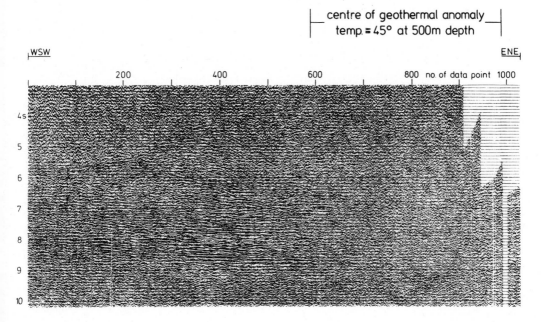

Fig. 5. 8-fold stacked and processed section of profile I.

thereby causing a tensional stress at lower crustal and upper mantle
levels. This is possibly the reason for the Tertiary volcanism in this
area. Towards the ENE the quality of the records deteriorates, most
probably because of strong noise sources in the vicinity of some cities.
Fig. 6 shows the same profile in form of a line-plot of clear reflections
without exaggeration, with the data from the long range refraction incorporated.

Velocity investigations so far are confined to the long profile I.
From the non-stacked sections with a common depth point arrangement along
the line first a rough stacking velocity was obtained from the curvature
of reflection. These values were used for first dynamical corrections which

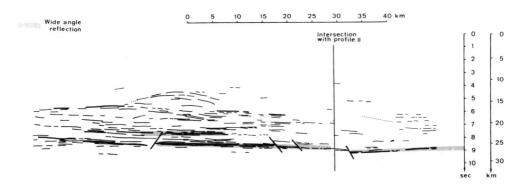

Fig. 6. Line plot of profile I; hatched area: refraction data.

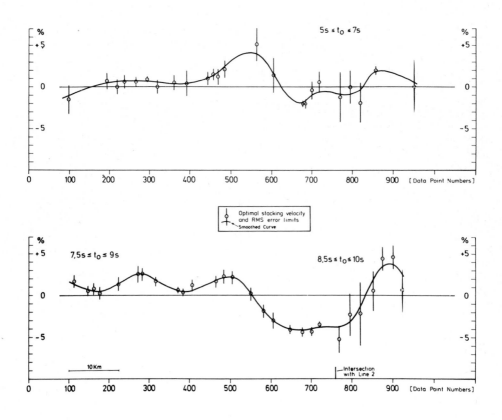

Fig. 7. Deviations from the average stacking velocities along profile I for two time intervals.

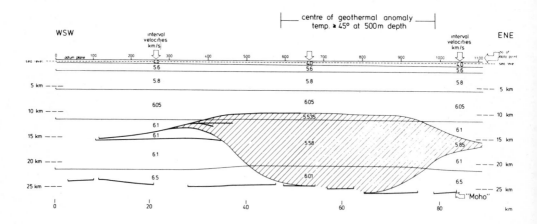

Fig. 8. Low velocity zone along profile I with corrected Moho depth and interval velocities in km/s.

later were improved by controlling the single-coverage sections along the
line. Stacking velocities were also obtained and controlled by applying
a number of different velocities for a trial stacking. The stacking
velocities were then linearly interpolated along the profile and used
for the dynamical corrections and the stacking of the data. There are
definitely deviations from the average stacking velocities along the profile. These deviations were mapped according to a method by Krey (1976).
Fig. 7 shows these lateral deviations of the stacking velocity along the
profile, separated for depths zones between 5 and 7 seconds and those
between 7.5 and 10 seconds two-way vertical travel time. The problem of
converting the observed lateral changes of stacking velocities into lateral
changes of average velocities was mathematically solved, applying certain
simplifying assumptions, by Lynn & Claerbout (1979). We attacked the
problem in a somewhat more rigorous manner by referring to a paper by
Hubral (1980). After some systematic trials and intermediate steps, always
controlling the observed stacking velocity with those calculated by the
model, the "final" model of Fig. 8 was obtained. In this model the "Moho"
shows some undulations. The general updip towards the WSW, i.e. towards
the Rhinegraben, is still there but much smaller than anticipated without
our knowledge about the strong velocity variations in the crust. It was
really a surprise that below Urach such a large low velocity body with up
to 10 % lateral velocity change exists. This is certainly much more than
a purely thermal effect of 100-200 °C could produce, and hydrothermal
alterations should be taken into consideration as well. The upper surface
of the low velocity zone coincides with a reflection of negative polarity
on the left hand side.

Conclusions

The method of calculating horizontal and vertical velocities along
extended reflection-refraction profiles provides a resolution of 1 % in
velocities. It is therefore well suitable for mapping those small velocity
variations which may be due to geothermal anomalies. For the center of
the Urach area velocity anomalies up to 10 % have been found at depths up
to 30 km in the crust. Also the lateral extension of this velocity anomaly
could be located. Moreover, the detailed 8-fold stacked reflection sections
show many items which may be strongly related to the geothermal anomaly,
such as a change of dip, indications of faults, inversion of reflection
polarity, and a relative maximum of crustal thickness in the Urach area.
    The high resolution in the determination of lateral velocities and the
unexpected strong deviations along the profile I imposes serious problems
and doubts on the accuracy of depth and structure determinations of
ordinary refraction work and teleseismic residuals.

References

Hubral, P., 1980: Wavefront curvatures in three-dimensional laterally
    inhomogeneous media with curved interfaces. - Geophysics 45 (5): 905-
    913.
Kern, H., 1978: The effect of high temperature and high confining pressure
    on compressional wave velocities in quartz-bearing and quartz-free
    igneous and metamorphic rocks. - Tectonophysics 44: 185-203.

Krey, T., 1976: Computation of interval velocities from common reflection point moveout times for layers with arbitrary dips and curvatures in three dimensions when assuming small shot-geophone distances. - Geophys. Prosp. 14: 91-111.

Lynn, W. & Claerbout, J., 1979: Velocity estimation in laterally varying media. - Paper pres. at the SEG 49th ann. internat. meeting, New Orleans.

Meissner, R., 1967: Zum Aufbau der Erdkruste. - Gerl. Beitr. z. Geophys. 76: 211-254 u. 295-314.

— , 1973: The Moho as a transition zone. - Geophys. Surv. 1: 195-216.

Meissner, R., Bartelsen, H., Krey, T. & Schmoll, J., 1980: Combined reflection and refraction measurements for investigating the geothermal anomaly of Urach. - In: A.S. Strub & P. Ungemach (Eds.), Advances in European Geothermal Research. - D. Reidel Publ. Comp., Dordrecht/ Netherl., 1086 pp.

Hot dry rock experiments in the Urach research borehole

H.-G. Dietrich, R. Haenel, G. Neth, K. Schädel and G. Zoth

with 8 figures

Dietrich, H.-G., Haenel, R., Neth, G., Schädel, K. & Zoth, G., 1982: Hot dry experiments in the Urach research borehole. - Geothermics and geothermal energy, eds. V. Čermák & R. Haenel, E. Schweizerbart'sche Verlagsbuchhandlung, Stuttgart: 293-299.

Abstract: The possibility of creating a heat extraction system with only one borehole has been investigated in the Urach research borehole. Fractures could be produced by hydraulic methods, and propping materials were utilized to keep them open. The breakdown pressure in the crystalline basement at 3300 m depth was about 450 bar; this was lower than expected (820 bar). This indicates a reopening of natural fractures. A circulation system for heat extraction could be set up.

Authors' addresses: H.-G. Dietrich, Stadtwerke Urach, 7417 Urach, Federal Republic of Germany; R. Haenel and G. Zoth, Niedersächsisches Landesamt für Bodenforschung, Stilleweg 2, 3000 Hannover, Federal Republic of Germany; G. Neth, Urbanstraße 53, 7000 Stuttgart, Federal Republic of Germany; K. Schädel, Geologisches Landesamt Baden-Württemberg, Albertstraße 5, 7800 Freiburg, Federal Republic of Germany

Introduction

Considerable efforts are being made to tap geothermal energy sources. One of these possibilities is heat extraction from hot dry rock.

This method has been developed and tested in Los Alamos, USA. Cold water is injected into a borehole. The water flows through an artificial fracture and is heated by the rock. The hot water or steam is then extracted through a second borehole and is available for utilization.

Within the scope of a research programme financed by the German Ministry for Research and Technology, the question of whether energy extraction is possible with only one borehole instead of two was to be tested.

The Urach research borehole was available for the investigations. Urach is located in the centre of a temperature anomaly south of Stuttgart in the Swabian Alb. The Urach research borehole reached the crystalline basement at a depth of about 1600 m. The drilling activities were stopped at 3334 m. The highest temperature measured at the deepest point of the borehole was about 143 °C.

Experiments

Two questions need to be answered before energy can be extracted from the crystalline basement. These constituted a significant part of the research programme:

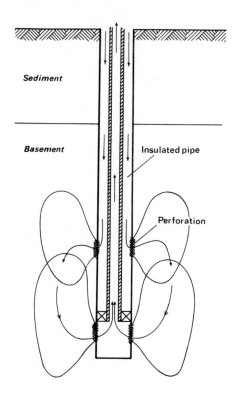

Fig. 1. A hypothetical basis for producing a water circulation (Preussag).

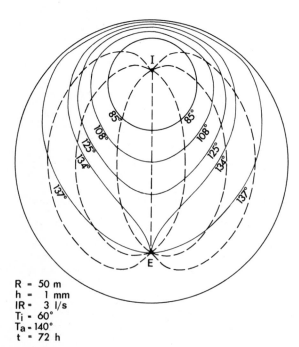

R = 50 m
h = 1 mm
IR = 3 l/s
T_i = 60°
T_a = 140°
t = 72 h

Fig. 2. Heat exchange process in the fracture between rock and water after Rodemann.

- Is it possible to create fractures in the basement in the present
  research borehole?
- Is it possible to circulate water between two fracs?

The so-called Preussag hypothesis (1977) was taken as the basis for the
following work, see Fig. 1. Cold water is injected through perforations
in the outer casing into the rock. The water is heated by the rock and
flows into the open space at the bottom of the borehole and through the
inner pipe to the earth's surface. The flow paths in the rock are natural
and artificially produced fissures. The heat exchange process in the
fractures between rock and water has been investigated theoretically by
Rodemann (1979). Fig. 2 shows a simplified model of the temperature
distribution and flow pattern (I = injection, E = extraction). The arrows
indicate the direction of flow.

On the basis of laboratory investigations, Rummel (1979) has calculated
the pressure necessary to create an artificial fracture hydraulically. The
results are shown in Fig. 3. Thus, at a depth of 3300 m at Urach, the
theoretical pressure is

$$P_f = 180 \text{ (bar)} + 294 \text{ (bar} \cdot \text{km}^{-1}) \cdot z \text{ (km)}$$

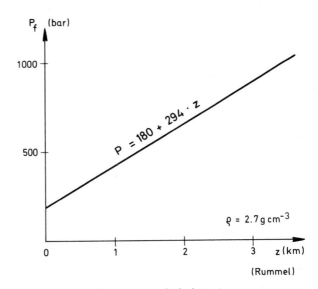

Fig. 3. Pressure curve necessary for creating an artificial fracture,
calculated for different depths on the basis of laboratory investigations.

Three curves showing the pressure in the rock during a frac are given in
Fig. 4. Curve 1 is typical of the ones obtained at Los Alamos, with a
peak indicating the breakdown pressure (arrow). Curves 2 and 3 show the
pressure within the Urach well during a frac in the cased part of the
borehole and in the uncased bottom part of the well, respectively. To
explain the discrepancy between the expected results and those which were
actually measured, it is supposed that natural fractures were reopened.
The lack of breaktops also shows that there is continuous opening of
existing fractures. This assumption is supported by the closed fractures
observed in the cores from the Urach well.

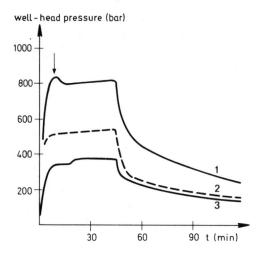

well - head pressure (bar)

Fig. 4. Pressure curves for the rock during the formation of fractures.

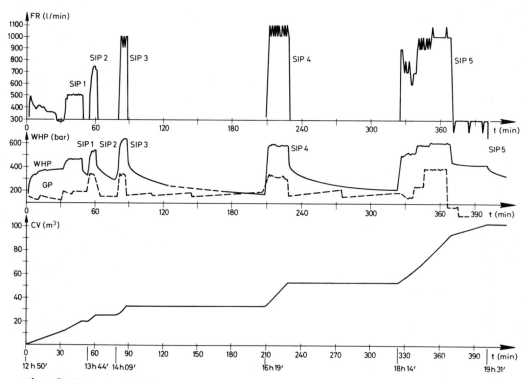

Fig. 5. Frac test 3.1.
1. Flow rate (FR) as a function of time;
2. Well-head pressure (WHP) as a function of time;
3. Cumulative volume (CV).
SIP = shut-in pressure, GP = guard pressure.

Four fracs were produced. One was created in the uncased bottom section of the well (Frac 1) and three in the cased part at 3260 m (Frac 2), 3280 m (Frac 3) and 3300 m (Frac 4). To do this, the casing was perforated at the corresponding depth over a length of about 5 m. It can be seen in Fig. 5 that flow rates of about 1200 1/min can be obtained when the pressure is relatively low, about 600 bar. During this investigation, 100 m³ of water was injected in 140 min with varying flow rates. This means that natural fractures in the crystalline basement at Urach can be easily re-opened and that they are widely branched.

With regard to the second objective, the fracs were treated to prepare them for the circulation of water. The fissures were propped open with "Superprop Norton-Props and 20/40 E 362". Versa-Gel was used as carrier to inject the Propsand through the perforations into the hydraulically opened fractures in the rock. This was done to lower the flow impedance during circulation of the water and prevent the closing of the frac.

Water was circulated between Frac 2 and Frac 3, and between Frac 3 and Frac 4 in the cased borehole, and between Frac 4 and Frac 1. The influence of pressure in Frac 2 on Frac 3 is shown in Fig. 6. The changes in pressure in Frac 2 (guard pressure) are shown in relationship to those in Frac 3 (well-head pressure). It can be seen that there is no influence when the difference in pressure is greater than 40 - 45 bar.

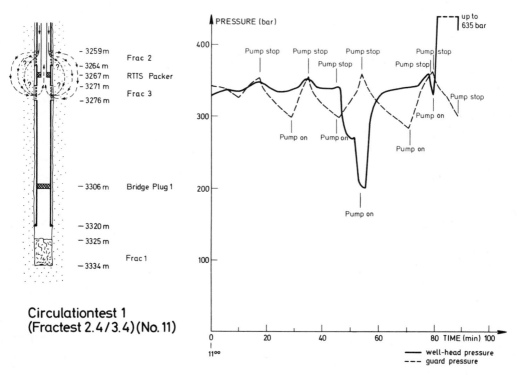

Fig. 6. Circulation test 1. There is no influence when the difference in pressure is greater than 40 - 45 bar.

One of the main objectives was achieved with the circulation tests between Frac 4 and Frac 1, e.g. circulation test 11, shown in Fig. 7. A nearly continuous flow rate of about 40 1/min at constant injection

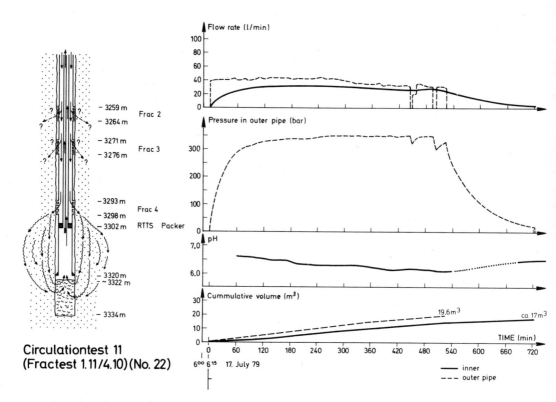

**Circulationtest 11
(Fractest 1.11/4.10) (No. 22)**

Fig. 7. Circulation test 11. Water circulation is possible at relatively low pressure.

pressure, with an immediate flow of water from the inner pipe was observed. A total of 19.6 m³ of water was injected and 17 m³ was extracted. During circulation the pH-value of the extracted water decreased from 6.7 to 6.2.

Two circulation tests were carried out in the cased borehole and ten tests between Frac 4 and Frac 1 at the bottom of the well. The vertical distance between the injection and extraction points of the water at the deepest point of the well was about 25 m, see Fig. 8. The lines indicate the inclination of the natural fractures in the rock samples from the depth where the fracs are produced.

Long-term circulation tests are planned as a next step. Heat exchange and temperature field will be determined. Therefore, the temperature difference between injection point and extraction point must be measured.

The results of this study will be compared with model calculations and should permit statements concerning the energy extraction over long periods of time. In addition, the chemical reaction of the water with the minerals in the fractures will be investigated with regard to impedance to the flow.

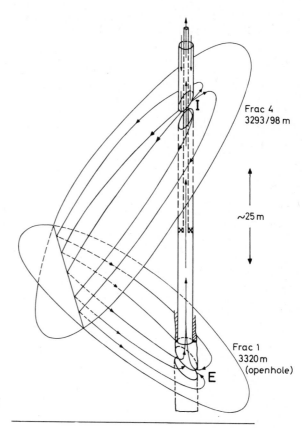

Fig. 8. Supposed circulation
system in the deepest part of
the borehole. I = injection,
E = extraction.

References

Ernst, P.L., 1977: A Hydraulic Fracturing Technique for Hot Dry Rock
    Experiments in a single borehole. - Society of Petroleum Eingineers of
    AIME, SPE 6897, Dallas, Texas.
Rodemann, H., 1978: Besuch bei der Hot-Dry-Rock-Arbeitsgruppe Los Alamos,
    USA. - Archive Nr. 81 653 NLfB-Hannover.
Rummel, F., 1979: Gesteinsphysikalische Daten, Bohrung Urach 3. - Report
    Az. 7084403 Ru/79-1, Ruhr Univ. Bochum.